# Environmental Issues
## Looking Towards a Sustainable Future

# Common Conversion Factors

**Length**
1 yard = 3 ft, 1 fathom = 6 ft

|  | in | ft | mi | cm | m | km |
|---|---|---|---|---|---|---|
| 1 inch (in) = | 1 | 0.083 | $1.58 \times 10^{-5}$ | 2.54 | 0.0254 | $2.54 \times 10^{-5}$ |
| 1 foot (ft) = | 12 | 1 | $1.89 \times 10^{-4}$ | 30.48 | 0.3048 | |
| 1 mile (mi) = | 63,360 | 5,280 | 1 | 160,934 | 1,609 | 1.609 |
| 1 centimeter (cm) = | 0.394 | 0.0328 | $6.2 \times 10^{-6}$ | 1 | 0.01 | $1.0 \times 10^{-5}$ |
| 1 meter (m) = | 39.37 | 3.281 | $6.2 \times 10^{-4}$ | 100 | 1 | 0.001 |
| 1 kilometer (km) = | 39,370 | 3,281 | 0.6214 | 100,000 | 1,000 | 1 |

**Area**
1 square mi = 640 acres, 1 acre = 43,650 $ft^2$ = 4046.86 $m^2$ = 0.4047 ha
1 ha = 10,000 $m^2$ = 2.471 acres

|  | $in^2$ | $ft^2$ | $mi^2$ | $cm^2$ | $m^2$ | $km^2$ |
|---|---|---|---|---|---|---|
| 1 $in^2$ = | 1 | | — | 6.4516 | — | — |
| 1 $ft^2$ = | 144 | 1 | — | 929 | 0.0929 | — |
| 1 $mi^2$ = | — | 27,878,400 | 1 | — | — | 2.590 |
| 1 $cm^2$ = | 0.155 | — | — | 1 | — | — |
| 1 $m^2$ = | 1,550 | 10.764 | — | 10,000 | 1 | — |
| 1 $km^2$ = | — | — | 0.3861 | — | 1,000,000 | 1 |

**Other Conversion Factors**

1 $ft^3$/sec = .0283 $m^3$/sec = 7.48 gal/sec = 28.32 liters/sec
1 acre-foot = 43,560 $ft^3$ = 1,233 $m^3$ = 325,829 gal
1 $m^3$/sec = 35.32 $ft^3$/sec
1 $ft^3$/sec for one day = 1.98 acre-feet
1 m/sec = 3.6 km/hr = 2.24 mi/hr
1 ft/sec = 0.682 mi/hr = 1.097 km/hr
1 billion gallons per day (bgd) = 3.785 million $m^3$ per day
1 atmosphere = $1.013 \times 10^5$ $N/m^2$ = approximately 1 bar
1 bar = approx. $10^5$ $N/m^2$ = $10^5$ pascal (Pa)

## Commonly Used Multiples of 10

| Prefix (Symbol) | Amount | Prefix (Symbol) | Amount |
|---|---|---|---|
| exa (E) | $10^{18}$ (million trillion) | centi (c) | $10^{-2}$ (one-hundredth) |
| peta (P) | $10^{15}$ (thousand trillion) | milli (m) | $10^{-3}$ (one-thousandth) |
| tera (T) | $10^{12}$ (trillion) | micro ($\mu$) | $10^{-6}$ (one-millionth) |
| giga (G) | $10^{9}$ (billion) | nano (n) | $10^{-9}$ (one-billionth) |
| mega (M) | $10^{6}$ (million) | pico (p) | $10^{-12}$ (one-trillionth) |
| kilo (k) | $10^{3}$ (thousand) | | |

## Volume

| | $in^3$ | $ft^3$ | $yd^3$ | $m^3$ | qt | liter | barrel | gal. (U.S.) |
|---|---|---|---|---|---|---|---|---|
| 1 $in^3$ = | 1 | — | — | — | — | 0.02 | — | — |
| 1 $ft^3$ = | 1,728 | 1 | — | .0283 | — | 28.3 | — | 7.480 |
| 1 $yd^3$ = | — | 27 | 1 | 0.76 | — | — | — | — |
| 1 $m^3$ = | 61,020 | 35.315 | 1.307 | 1 | — | 1,000 | — | — |
| 1 quart (qt) = | — | — | — | — | 1 | 0.95 | — | 0.25 |
| 1 liter (l) = | 61.02 | — | — | — | 1.06 | 1 | — | 0.2642 |
| 1 barrel (oil) = | — | — | — | — | 168 | 159.6 | 1 | 42 |
| 1 gallon (U.S.) = | 231 | 0.13 | — | — | 4 | 3.785 | 0.02 | 1 |

## Energy and Power

1 kilowatt-hour = 3,413 Btus = 860,421 calories

1 Btu = 0.000293 kilowatt-hour = 252 calories = 1,055 joule

1 watt = 3.413 Btu/hr = 14.34 calories/min

1 calorie = the amount of heat necessary to raise the temperature of 1 gram (1 $cm^3$) of water 1 degree Celsius

1 quadrillion Btu = (approximately) 1 exajoule

1 joule = 0.239 calorie = $2.778 \times 10^{-7}$ kilowatt-hour

## Mass and Weight

1 pound = 453.6 grams = 0.4536 kilogram = 16 ounces

1 gram = 0.0353 ounce = 0.0022 pound

1 short ton = 2,000 pounds = 907.2 kilograms

1 long ton = 2,240 pounds = 1,008 kilograms

1 metric ton = 2,205 pounds = 1,000 kilograms

1 kilogram = 2.205 pounds

# Environmental Issues
## Looking Towards a Sustainable Future

Fourth Edition

**Robert L. McConnell, Ph.D.**
Senior Fellow
U.S. Partnership for Education for Sustainable Development
and Professor Emeritus, Environmental Science and Geology,
University of Mary Washington, Fredericksburg, Virginia
Present address: Corvallis, OR

**Daniel C. Abel, Ph.D.**
Senior Fellow
U.S. Partnership for Education for Sustainable Development
and Professor, Department of Marine Science,
Coastal Carolina University, Conway, South Carolina

Cover Art: Courtesy of Daniel C. Abel

Pearson Learning Solutions, 501 Boylston Street, Suite 900, Boston, MA 02116
A Pearson Education Company
www.pearsoned.com

Printed in the United States of America

2 3 4 5 6 7 8 9 10 V0CR 17 16 15 14 13


000200010271724642

MC/CM

ISBN 10: 1-256-93309-0
ISBN 13: 978-1-256-93309-0

# CONTENTS

Preface   v

## Basic Concepts and Tools   1

## Part One ✪ Principles of Sustainability   17

## Part Two ✪ Population and Migration   29

1   Global Population Growth: *Is it Sustainable?*   29
2   Carrying Capacity and Ecological Footprint   37
3   Coastal Population Growth: Bangladesh   42
4   Population Growth and Migration   48

## Part Three ✪ Climate Change   53

5   Greenhouse Gases and Climate Change: Part One   53
6   Greenhouse Gases and Climate Change: Part Two   60

## Part Four ✪ Energy   69

7   Oil and Natural Gas   69
8   Coal   81
9   Bringing the World to the U.S. Standard of Living   90
10   Sustainable Energy: *Is the Answer Blowing in the Wind?*   95

## Part Five ✪ Consumption and the Quality of Life   105

11   Global Water Supplies: *Are They Sustainable?*   105
12   Motor Vehicles and the Environment I   118
13   Motor Vehicles and the Environment II: *Global Trends*   125
14   Whacker Madness? *The Proliferation of Turfgrass*   132
15   Mountains of Trash: *Are They Sustainable?*   144
16   Gold Mining: *Is It Sustainable?*   156
17   Persistent Organic Pollutants (POPs)   166

## Part Six ✪ Threats to Ecosystems   175

18   Global Grain Production: *Can We Beef It Up?*   175
19   Soils and Sustainable Societies   184

| | | |
|---|---|---:|
| 20 | The State of Global Forests | 195 |
| 21 | Restoring Estuaries: *Chesapeake Bay* | 208 |
| 22 | Illegal Immigration: *Ballast Water and Exotic Species* | 217 |
| 23 | Catch of the Day: *The State of Global Fisheries* | 228 |

## Part Seven ◉ Sustainability and the Individual — 239

| | | |
|---|---|---:|
| 24 | Sustainable Communities: *Sprawl versus Smart Growth* | 239 |
| 25 | Sustainable Coastal Development | 247 |
| 26 | Sustainable Buildings and Housing | 254 |
| 27 | The Three R's: Reduction, Reuse, and Recycling | 265 |
| 28 | A Sustainable Diet | 277 |
| 29 | The Sustainable Campus | 287 |
| 30 | Restoration Ecology | 297 |

| | |
|---|---:|
| Index | 305 |

# PREFACE

## TO THE INSTRUCTOR

The idea for this book arose when we were colleagues at the University of Mary Washington. We grew impatient with a teaching style centered on the faculty member as "lecturer" and expert and the student as "scribe" and novice. We felt that such an approach encourages students to be passive rather than active learners and leads to an unhealthy dependency on the faculty person as "expert."

We also found that many students were afraid of math, were rusty in its use, or were superficially trained in arcane fields of calculus. This lack of math skills often leaves students unprepared to deal with the complexity of today's environmental issues. Moreover, we are continually surprised to discover how many bright students can't do three things: understand and USE the units of the metric system, use scientific notation, and critically evaluate complex environmental issues.

Most of a student's discomfort with math is generally founded in frustration. For example, making one error in a series of calculations can render the whole effort useless. We believe that, in the absence of a real learning disability, to solve most math problems requires no special aptitude, only clear sequential instructions and attention to detail. That is why step-by-step calculations are included in your Answer Key.

One of our major objectives is thus to help develop math literacy (*numeracy*) among today's students. We understand that many students have some "math anxiety," so we have included a section entitled "Using Math in Environmental Issues," in which we use a step-by-step method to take students through examples of the calculations in the Issues. We even show sample keystrokes involved in using common calculators. We believe this method will gradually build the student's confidence enough to trust his or her own efforts. Math proficiency is one of the important skills necessary for fully understanding environmental issues, and without these skills, the student's only option is to make choices on the basis of which "expert" is most "believable." Such skills involve the ability to manipulate large numbers using scientific notation and exponents, the ability to use compound growth equations containing natural logs, and so on.

It is our goal that this book be provocative, factually accurate, and up-to-date. *Environmental Issues: An Introduction to Sustainability* is meant be the basis for an issues-oriented introductory, seminar, upper-level, or laboratory course in environmental science or studies. In addition, this newly revised fourth edition is appropriate as a stand-alone text for courses in sustainability or sustainable development. However, it can also be used as a supplement to traditional texts in environmental science, geology, biology, and other natural sciences. It is also suitable for humanities courses that seek to cultivate an awareness of and knowledge about environmental issues and sustainability. You can use as projects to develop students' critical thinking skills in a deliberate and structured way. By their nature, they require students to integrate topics from across subdisciplines to measure, analyze, and evaluate each issue using the discipline and method of a scientist.

But becoming educated is much more than simply acquiring skills. Therefore, we have two additional objectives: to provide students with the knowledge and intellectual standards necessary to apply critical thinking to environmental studies and sustainability, and to foster their ability to critically evaluate issues.

As such, *Environmental Issues: An Introduction to Sustainability* is as much an *interactive* workbook as a traditional textbook. We expect students to have access to standard references in environmental, physical, and natural sciences and to have access to and know how to use the Internet. Indeed, every Issue contains URLs (uniform resource locators) to websites for up-to-the-minute information.

We also trust that you, as an expert in your field and with your own perspectives, will supplement the information in this book with your own comments, introductions, and critical comments on the questions we ask your students. The issues are user-friendly and based on solid science. We define key terms, and keep jargon to a minimum. When important terms are introduced they are italicized and defined if necessary. We introduce and explain key mathematical formulas using a step-by-step nonthreatening approach that we hope you will appreciate.

As we have mentioned, one of our major objectives is to foster proficiency among today's students in the kind of math they need to properly quantify environmental issues, such as the use of key formulas, scientific notation, and the metric system. We provide detailed introductions for each of these topics, as well as a detailed Answer Key for you to use as you see fit that shows the step-by-step calculations used to determine the answers. We have purposefully not provided direct student access to the Answer Key. In addition, we provide you with suggested answers to the "Critical Thinking" questions, but we are confident that you will have your own point of view that you will wish to develop for many if not most of them.

To encourage rigorous critical thinking, there are questions with spaces for answers integrated into each issue. Critical thinking implies using a set of criteria and standards by which the reasoner constantly assesses his or her thinking. At the core of critical thinking is self-assessment.

We devote a detailed section to "critical thinking" in the first section of this book. We think *it is vitally important that students read this material for content, as they will be asked to apply these standards and criteria throughout the book.*

Many questions have been designed to be provocative. This may lead you or your students to perceive a "bias" in the wording of some of the questions. Although we have made the content as factual as possible, we do have strong convictions about these issues. *Convictions are not, however, biases.* Based on the scientific method, our views as scientists are subject to change as evidence supporting our convictions changes. And as such, we are constantly testing the *assumptions* that we use when approaching complex environmental issues. Indeed, this aspect can be turned to a major advantage. Ask your students to look for examples of bias in the questions, and then discuss with them the difference in science between "bias" and "conviction." No doubt it will prove a fruitful activity and may lead students into research (perhaps to "prove us wrong"), which is the essence of progress in the search for scientific truth.

## New Features in the Fourth Edition

All Issues in the fourth edition are either updated and rewritten versions of Issues in the third edition, or contain new content. The third edition's simple, straightforward format and black-and-white figures have been retained. In recognition of the exciting and important developments in the field of sustainability, we have integrated information and questions on sustainability throughout the text, and hope that this book will be useful to the new generation of sustainability and sustainable development courses springing up on campuses worldwide. Although the section "Sustainability and the Individual" follows the first twenty-three Issues of the book, we encourage you to take the time to do one or more of these Issues, which we wrote to serve as capstone studies. As in the third edition, each of the Issues is designed to stand-alone, and you can do them in virtually any order. However, we feel Issue 1 is pivotal, and it should be completed first and very carefully.

We apologize in advance for any typos and errors we did not catch.

## How to Use This Book Effectively

Here's how we have used *Environmental Issues* in our classes. First, we typically dedicate class time or a portion of a laboratory early in the semester to introducing students to the principles of critical thinking. In this period we ask students to list characteristics of critical thinking (i.e., higher order thinking, or just plain good, effective thinking). More often than not, the class's list encompasses many of the standards of critical thinking that are contained in the section of this book entitled Basic Concepts and Tools: Using Math and Critical Thinking. As we go over these characteristics, we emphasize *clarity, awareness of assumptions,* and *continuous self-assessment*, as well as the importance of applying critical thinking to environmental (and other) issues. We then analyze passages from letters to the editor, newspaper op-eds, and popular magazine articles.

It is also worthwhile, before using this book, to confront students' math anxiety early and attempt to reassure them that they are capable of doing the math, although they may be rusty and require some practice and assistance.

We frequently assign Issues or blocks of Issues as group projects, in which students collaborate using a webpage, "blackboard," or other form of electronic discussion group, and we try to ensure that every student logs on to the system. They can exchange information, send reports, references, and less-threatening critiques to classmates, and so on. We recommend you encourage students to do this, while reminding them that few people are won over to another's argument by having their own ideas ridiculed.

For classes of more than twenty-four students, use of a group e-mail system or other similar classroom software package offers you enormous possibilities. You can break the class into working groups of three to four members, and the group can be reorganized as the term progresses. Working groups can do the calculations independently, can check their work by e-mail or in meetings, and can get together to hash out the answers to the "For Further Thought" questions.

This method also allows you to monitor the class's work and communicate with the class in a nonthreatening manner. Your institution's computer or information technology services department can provide you with the details if you've never used one.

We have found that many students who would be reluctant to participate in a classroom discussion will willingly contribute in the relative anonymity of an electronic discussion group. Students can (and probably should) copy their math and send it to others over the system, thereby checking each other's work. You can send comments to the groups if you feel they are on the wrong track and encourage them if they are making progress. You can send them questions arising out of their own discussions and respond immediately to their inquiries. They can exchange information on the "For Further Thought" questions, and you can discuss these questions with the students as their work evolves.

Students can debate the issues and grade each other, or they can turn in their work in the normal fashion and be graded on the accuracy of their calculations as well as on the thoroughness of their answers.

Work done on these issues can take the place of one or more exams or quizzes, freeing you for other activities and providing students with a less threatening way to earn class credit than an all-exam format. We believe that students retain more information from work on projects and reports than from cramming for tests. At the end of the term, you could pass out a study sheet detailing specifically what material or questions from the Issues will be covered on the final exam, if you choose to test them on their work in this manner.

You may contact us at rmcconne@umw.edu and dabel@coastal.edu with questions and comments.

# TO THE STUDENT

If you are concerned about and interested in environmental issues but feel you just "can't deal with the math," then this book is for you. As environmental scientists, we care deeply about environmental issues, and we feel that you, as a responsible citizen who will have to make increasingly difficult choices in the years ahead, need to be concerned about them as well. Here are a few of our reasons:

- A STRONG scientific consensus exists that our growing human population is having a measurable and harmful effect on the composition of the planet's atmosphere. The world's leaders continue a dialogue on how to respond to human-induced global climate destabilization, a result of global warming. Even though we can't yet be certain what the scale of these changes will be, evidence suggests that the impact will, on the whole, be negative, and could be catastrophic for hundreds of millions of people crowded into many of the planet's coastal cities.
- Marine scientists are concerned about the very survival of many marine organisms, including some that form the basis of major world fisheries. In fact, entire ecosystems such as coral reefs may be at risk. The ocean's ability to absorb our waste, toxic and otherwise, is certainly limited, and these limits have likely been exceeded.
- Although we in the United States and in a few other areas of the world have made impressive strides in improving or at least slowing the degradation of our air, water, and soil, our relentlessly growing human numbers threaten this progress.
- Growing levels of material consumption in developed countries coupled with less-regulated international commerce (free trade) are placing increased stress on critical ecosystems such as tropical and boreal forests, in turn gravely threatening the planet's species diversity.
- Environmental issues, such as water conflicts in the Middle East and the trans-national impact of air and water pollution could destabilize international relations and lead to regional conflicts.
- Fossil fuels continue to be the basis of industrial and postindustrial society. Their extraction, transportation, and use impose significant costs on the planet. Much of this cost is externalized (or dumped) onto the environment. As the 2010 Deepwater Horizon Platform incident in the Gulf of Mexico illustrated, addressing environmental issues should involve an inclusion of these costs.

We hope you will find this book to be a provocative introduction to a number of these issues and to many others that you may have never even thought about. These are real-life issues, not hypothetical ones, and you need certain basic skills to fully understand them.

- You must be familiar with the units of the metric system.
- You must be able to use a few mathematical formulas to quantify the issues you will be debating, and you must be able to carry out the calculations accurately.
- We show you how to do this.

- You should develop the habit of rigorously assessing your thinking, and you should apply certain critical thinking skills and techniques when discussing the implications of your calculations.
- You should realize that critical thinking requires practice, dedication, time, and an open mind.

We use a step-by-step method to take you through many of the calculations in this book. Math proficiency is one of the important skills necessary for fully understanding environmental issues, and without these skills, your only option is to make choices on the basis of which "expert" you believe.

But becoming educated is much more than acquiring skills. Therefore, we have two further objectives: to provide you with the knowledge and intellectual standards necessary to apply critical thinking to environmental studies, and to foster your ability to critically evaluate issues, a skill without which mechanical skills are of limited value. To that end, we have included a section called "Basic Concepts and Tools: Using Math and Critical Thinking," in which we illustrate and describe each criterion for critical thinking, as well as the framework within which these criteria are applied. It is very important that you read this section carefully and do all of the exercises in it *before* you begin the analyses of the issues.

Let's consider an example: growth. The word is used in many societal, economic, demographic, and environmental contexts, including growth of the economy, growth of the population, growth of impervious surfaces, growth of food production, and growth of energy use. Assessing the impact of growth requires an understanding of a few simple equations such as the compound interest equation. You should be able to use it accurately and understand its implications. We show you how.

As our national and global population grows and changes and our relationships with other nations and peoples evolve, environmental issues will become even more complicated. Domestically, demographic and ethnic changes are becoming more important, which in turn requires an enhanced ability to critically evaluate issues.

We don't try to avoid controversial topics such as population growth, personal consumption, the automobile, and immigration.

We hope you will be challenged by the issues discussed in this text and that you will research them and become an "expert" on the topics yourself. In fact, if we may be allowed a hidden agenda, this is it.

And while you are testing your own thinking and reasoning, test ours as well. Look for examples of "bias" in the "Critical Thinking" questions, and be prepared to support your conclusions and discuss them with your fellow students and with your instructor.

## A Note on Conventions Used in This Book

When we express rates, concentrations, and so on—for example, milligrams per liter, people per hectare, tonnes per year—we use all acceptable formats, including "per" (as above), the slash (mg/L, people/hectare, tonnes/year), and at times negative exponents ($mg \cdot L^{-1}$, people $\cdot$ hectare$^{-1}$, tonnes $\cdot$ year$^{-1}$).

## A Word about Calculators

A handheld calculator can make analyzing environmental issues easier, or it can be a source of great frustration. It all depends on how carefully you put the information into the calculator and how familiar you are with how to use it.

We recommend buying the simplest calculator you can find that will carry out the tasks you need performed. These are typically sold as "scientific calculators." *We also encourage you to take the time to learn how to use the calculator properly.* The calculator should perform the basic data manipulation functions—such as adding, subtracting, multiplying, dividing, determining squares and square roots—and have a single, rather than multiple, memory that is easy to use. Some additional features you should look for include the following:

- Parentheses () keys
- A $y^x$ key
- A reciprocal (**1/x**) key
- Ability to do simple statistics, including means (**x**) and standard deviations (**S**)
- An **ex** key
- A **LN** (natural log) key
- An **exp** or **EE** key

## ACKNOWLEDGMENTS

This book could not have succeeded without the aid of numerous friends and colleagues. We appreciate the invaluable feedback from the following reviewers of previous editions, whose comments and suggestions materially improved the manuscript: Janet Kotash, Moraine Valley Community College; Debra Rowe, Oakland Community College; Ravi Srinivas, University of St. Thomas; Daniel J. Sherman, University of Puget Sound; and Maud Walsh, Louisiana State University. Special thanks to Coastal Carolina University students Lyndsey King and Chelsea Norman, who helped edit the 4th edition of this book.

**Robert L. McConnell**
**Daniel C. Abel**

*Rather than to travel into the sticky abyss of statistics, it is better to rely on a few data and on the pristine simplicity of elementary mathematics.*
—ALBERT BARTLETT


*An unexamined life is not worth living.*
—SOCRATES

# BASIC CONCEPTS AND TOOLS: USING MATH AND CRITICAL THINKING

Here we review the units of the metric system and the rules for using scientific notation in math problems. We will also show you how to do more complicated math, such as projecting population based on growth rates, using step-by-step methods. This chapter also deals with *critical thinking*. At first glance many issues seem perplexing and hard to approach. Don't panic; we will show you how to apply intellectual standards within a critical thinking framework.

## THE METRIC SYSTEM

The metric system's elegance and utility arise from its simplicity. You probably already know that the metric system is based on powers of ten: 10 millimeters in 1 centimeter, 100 centimeters in 1 meter, and so forth. This key point should be kept in mind.

You must be able to make certain conversions from the English system to the metric system and back. (It is a good idea to mark this page, since you will find it useful to refer to again.) Some common conversions as well as metric prefixes are given in Tables 1 through 4 below and may also be found inside the front cover of this book.

Here's a useful shortcut. To convert areas, you can use the conversion factors for the units of length and square them. For example, to convert 6 square feet to square yards, do the following:

$$6 \text{ ft}^2 \times (1 \text{ yd/3 ft})^2 = 6 \text{ ft}^2 \times 1 \text{ yd}^2/9 \text{ ft}^2 = 0.67 \text{ yd}^2$$

We will introduce and explain other units of the metric system as the need arises. One relationship you will find especially helpful: There are 1,000 liters per cubic meter of water under standard conditions.

Now do these self-assessment questions to test your ability to manipulate and convert these units. These skills are essential for the work in the issues that follow. Remember to use your conversion factors (so that units cancel each other out). For example, to determine how many liters are in 3.6 cubic meters:

$$3.6 \text{ m}^3 \times 1000 \text{ L/m}^3 = 3600 \text{ L}$$

**TABLE 1** ■ Units of the Metric System

| Units of Distance: The fundamental unit is the *meter* | |
|---|---|
| 1000 ($10^3$) m = 1 km | One thousand meters = one kilometer |
| 100 ($10^2$) cm = 1 m | One hundred centimeters = one meter |
| 10 ($10^1$) mm = 1 cm | Ten millimeters = one centimeter |
| 1000 ($10^3$) $\mu$m = 1 mm | One thousand micrometers = one millimeter |
| **Units of Mass: The fundamental unit is the *gram*** | |
| 1000 kg = 1 metric ton | One thousand kilograms = one (metric) tonne |
| 1000 g = 1 kg | One thousand grams = one kilogram |
| 1000 mg = 1 g | One thousand milligrams = one gram |
| 1,000,000 ($10^6$) ng = 1 mg | One million nanograms = one milligram |
| **Units of Volume: The fundamental unit is the *liter*** | |
| 1000 L = 1 $m^3$ | One thousand liters = 1 cubic meter |
| 1000 ml = 1 L | One thousand milliliters = 1 liter |

**TABLE 2** ■ Metric Prefixes and Equivalents

| Large Numbers | | |
|---|---|---|
| One thousand | =1000 | =$10^3$ (kilo or k) |
| One million | =1,000,000 | =$10^6$ (mega or M) |
| One billion | =1,000,000,000 | =$10^9$ |
| One trillion | =1,000,000,000,000 | =$10^{12}$ |
| One quadrillion | =1,000,000,000,000,000 | =$10^{15}$ (commonly used in expressions of energy use) |
| **Small Numbers** | | |
| One hundredth | =1/100 | =$10^{-2}$ (centi or c) |
| One thousandth | =1/1000 | =$10^{-3}$ (milli or m) |
| One millionth | =1/1,000,000 | =$10^{-6}$ (micro or mc or $\mu$) |
| One billionth | =1/1,000,000,000 | =$10^{-9}$ (nano or n) |

**TABLE 3** ■ Some Common Metric Conversions

| gallons/liters | 1 U.S. gal. = 3.8 L | One U.S. gallon = 3.8 liters |
|---|---|---|
| liters/gallons | 1 L = 0.264 U.S. gal. | One litre = 0.264 U.S. gallon |
| meters/yards | 1 m = 1.094 yd | One meter = 1.094 yards |
| yards/meters | 1 yd = 0.914 m | One yard = 0.914 meter |
| grams/ounces | 1 g = 0.035 oz | One gram = 0.035 ounce |
| ounces/grams | 1 oz = 28.35 g | One ounce = 28.35 grams |
| kilograms/pounds | 1 kg = 2.2 lb | One kilogram = 2.2 pounds |
| pounds/grams | 1 lb = 454 g | One pound = 454 grams |
| miles/kilometers | 1 mi = 1.609 km | One mile = 1.609 kilometers |
| kilometers/miles | 1 km = 0.621 mi | One kilometer = 0.621 mile |

**TABLE 4** ■ Conversion Factors for Area

| | | |
|---|---|---|
| square miles/square kilometers | $1 \text{ mi}^2 = 2.6 \text{ km}^2$ | One square mile = 2.6 square kilometers |
| square kilometers/square miles | $1 \text{ km}^2 = 0.39 \text{ mi}^2$ | One square kilometer = 0.39 square mile |
| hectares/acres | 1 ha = 2.47 acres | One hectare = 2.47 acres |
| acres/hectares | 1 acre = 0.4 ha | One acre = 0.4 hectare |
| square yards/square meters | $1 \text{ yd}^2 = 0.84 \text{ m}^2$ | One square yard = 0.84 square meter |
| square meters/square yards | $1 \text{ m}^2 = 1.2 \text{ yd}^2$ | One square meter = 1.2 square yards |
| square miles/acres | $1 \text{ mi}^2 = 640$ acres | One square mile = 640 acres |

**Question 1:**   How many micrometers are in a meter?

**Question 2:**   How many centimeters are in a kilometer?

**Question 3:**   How many grams are in a tonne? (*Tonne* is the correct spelling for the metric unit of 1000 kg.)

**Question 4:**   Express your height in feet, meters, and centimeters.

**Question 5:**   Express your weight in kilograms and pounds.

## SCIENTIFIC NOTATION

Very large numbers are most conveniently manipulated (i.e., added, subtracted, multiplied, and divided) by converting the numbers to logarithms to the base 10. Since this is not a math book, we are not going to delve into the theory of logarithms; a practical application is all you need.

*The basic fact you need to know is that 100 (pronounced "ten to the zero" or "ten to the zero power") is defined as 1.*

The skills to do the manipulations are easy to learn. The first step is to convert large numbers to scientific notation, with which you should already be familiar. For example, in scientific notation, 18,000,000 is $1.8 \times 10^7$.

Try another example.

**Question 6:** Express one billion (1,000,000,000) in scientific notation.

Now let's introduce a wrinkle.

**Question 7:** Express 2,360,000 in scientific notation.

You can express the same number in a variety of ways using exponents, such as $23.6 \times 10^5$, and they all will mean the same thing. But it is customary to express all values in the same format, typically by placing only one digit to the left of the decimal place (e.g., $2.36 \times 10^6$).

**Question 8:** Express 23,000,000,000,000 (23 trillion) the customary way using exponents.

For numbers summarized in scientific notation, the prefix in front of the unit (e.g., kilo) denotes the magnitude of the unit (e.g., *kilo*grams = units of 1000 g).

**Question 9:** Convert 1.86 mm to (a) nm, (b) mm (micrometers), (c) cm, (d) m, and (e) km. Express your answers as decimals and in scientific notation.

## MANIPULATING NUMBERS EXPRESSED IN SCIENTIFIC NOTATION

To add, subtract, multiply, and divide large numbers using scientific notation, you need to remember a few basic rules and check your work carefully. That is all there is to it. Here are the rules.

## Multiplication Using Scientific Notation

To multiply numbers expressed in scientific notation, *multiply the bases and add the exponents*. For example, to multiply

$$(3 \times 10^3) \times (4 \times 10^5)$$

multiply the bases

$$3 \times 4 = 12$$

and add the exponents

$$3 + 5 = 8$$

The result is $12 \times 10^8$ or, using the appropriate convention, $1.2 \times 10^9$. (NOTE: This is the same as $120 \times 10^7$ and any number of other variations as well.)

## Division Using Scientific Notation

To divide numbers expressed in scientific notation, *divide the number in the numerator by the number in the denominator and subtract the exponent of the denominator from the exponent of the numerator*. Recall that:

$$\frac{\text{NUMERATOR}}{\text{DENOMINATOR}}$$

For example, to divide

$$(5.2 \times 10^4) \div (2.6 \times 10^2)$$

divide the numerator by the denominator

$$5.2 \div 2.6 = 2$$

and subtract the exponent of the denominator from the exponent of the numerator

$$4 - 2 = 2$$

So the answer is $2 \times 10^2$.

**Question 10:** Perform the following manipulations:

$$(8.7 \times 10^{-3}) \times (4.2 \times 10^{-9}) = \underline{\hspace{3cm}}$$

$$(5.2 \times 10^{18}) \times (8.7 \times 10^{22}) = \underline{\hspace{3cm}}$$

$$(8.7 \times 10^{-3}) \div (4.2 \times 10^{-9}) = \underline{\hspace{3cm}}$$

$$(5.2 \times 10^{18}) \div (8.7 \times 10^{22}) = \underline{\hspace{3cm}}$$

## Addition Using Scientific Notation

To add numbers expressed in scientific notation, *you simply add the numbers after converting both to the same exponent.*

For example, to add 3 billion to 14 million,

$$3,000,000,000 + 14,000,000$$

or

$$(3 \times 10^9) + (14 \times 10^6)$$

convert both numbers to the same exponent (it doesn't matter which, but it is usually easier to use the smaller):

$$3 \text{ billion} = 3000 \text{ million or } 3000 \times 10^6$$
$$14 \text{ million} = 14 \times 10^6$$

Therefore,

$$(3000 \times 10^6) + (14 \times 10^6) = 3014 \times 10^6$$

An equally correct answer would be $3.014 \times 10^9$ which is the customary way to express the answer.

Now work out the rule for subtraction using scientific notation.

## USING MATH IN ENVIRONMENTAL ISSUES

The following is an introduction to some of the formulas used in this book.

### How to Project Population Growth Using the Compound Growth Equation

For this you will need a calculator with an exponent key. The equation is

$$\textbf{future value} = \textbf{present (or starting) value} \times \textbf{(e)}^{\textbf{rt}}$$

where e equals the constant 2.71828 . . . , r equals the rate of increase (expressed as a decimal, e.g., 5% would be 0.05), and t is the number of years over which the growth is to be measured.

Replacing words with symbols, this equation becomes

$$\mathbf{N = N_0 \times (e)^{rt}}$$

The variable $N_0$ represents the value of the quantity at time zero, that is, the starting point.

This equation is central to understanding exponential growth and is one of the few worth memorizing. Using it is not as intimidating as you may think.

***Sample Growth Calculation.*** Here's an example. Let's figure out (demographers use the word *project*) the world population in 2020, given the mid-year 2006 population of 6.52 billion ($6.52 \times 10^9$ or 6,520,000,000) and a growth rate of 1.14% per year. You can obtain current monthly world population figures at the Census Bureau's World POPClock website (www.census.gov/cgi-bin/ipc/popclockw).

Here's how to do this calculation:

$$2020 \text{ population} = (6.52 \times 10^9) \times e^{(0.0114 \times 14)}$$

On a typical, nongraphics calculator, keystrokes are as follows (commas are for punctuation only):

Key in **0.0114** (the decimal equivalent of 1.14%), then × (multiply sign), then **14,** then **= .** This gives you the exponent (the number to which e must be raised). Next hit the button labeled $\mathbf{e^x}$. Note that on most calculators, $e^x$ is labeled *above* a button having another label (frequently **ln**). If this is the case on your calculator, you must hit the key labeled **2nd** or

**2nd F** first, followed by the key with $e^x$ above it. Further, some calculators require you to hit the = key after the $e^x$ key.

Next, key in × (multiply sign), followed by the *starting value* ($6.52 \times 10^9$). On most calculators this is done by keying in **6.52,** then hitting the button labeled **EE** or **EXP,** then keying in **9.** If the **EE** or **EXP** label is above the key, recall that you must first hit the key labeled **2nd** or **2nd F** before punching **EE** or **EXP.** Finally, hit the = sign.

The correct answer is 7.65 billion.

Now, project the population of Dhaka, Bangladesh, in 2056 at a growth rate of 4.2% per year using the compound growth equation. The 2005 population was 12.6 million.

$$\text{2056 population} = (12.6 \times 10^6) \times e^{(0.042 \times 51)}$$

$$= 1.07 \times 10^8$$

$$= 107 \times 10^6$$

$$= 107 \text{ million}$$

The compound growth equation can also be rearranged. If you know the starting and ending population sizes over a given period, you can calculate the average growth rate over that period using the formula

$$r = (1/t) \ln(N/N_0)$$

Also, you can calculate how long it would take a population of a given size to grow (or decrease) to a different size at a specified growth rate using

$$t = (1/r) \ln(N/N_0)$$

## Doubling Time

When a population grows exponentially (by a percentage of the original number), the time it takes for the population to double, called *doubling time* (symbol "**t**"), can be approximately calculated using the following formula:

$$t = (70/r)$$

where t is the doubling time (usually in years) and r is the growth rate expressed as the decimal increase or decrease × 100 (for example, you would enter 7 for a 7% increase).

***Derivation of Doubling Time.*** To derive the doubling time formula, we revisit the compound growth equation:

$$\text{future value} = \text{present value} \times (e)^{rt}$$

where e equals the constant 2.71828..., r equals the rate of increase (expressed as a decimal, i.e., 5% would be 0.05), and t is the number of years (or hours, days, etc.—whatever units you are using in r) over which the growth is to be measured.

Replacing words with symbols, this equation becomes:

$$N = N_0 \times (e)^{rt}$$

The variable $N_0$ represents the value of the quantity at time zero, that is, the starting point.

If a population doubles in size (that is, increases by a factor of 2), the ratio $N/N_0$ would be exactly 2.

The equation is thus rearranged as follows:

$$N/N_0 = e^{rt}$$
$$2 = e^{rt}$$

Taking the natural log of each side of the equation, we get

$$\ln 2 = \ln(e^{rt})$$
$$\ln 2 = rt$$
$$\ln 2 = 0.693$$
$$0.693 = rt$$

Dividing by r:

$$0.693/r = t$$

For convenience sake, we round 0.693 to 0.70, and we also multiply the left side of the equation by 100/100, which allows us to enter the rate as a percentage (i.e., 5% would now be entered as 5 instead of 0.05). Thus the final doubling time formula:

$$t = 70/r$$

# CRITICAL THINKING[1]

"It is a humanist principle that if you want to know the truth, go to the sources, not to the commentators."[2]

## Overview

The mind has awesome power. John Milton said,

> The mind is its own place,
> And in itself can make
> A Heav'n of Hell,
> A Hell of Heav'n.

We must use the mind's power effectively, which involves critical thinking. Critical thinking is sometimes called *second order thinking*. *First order thinking* is spontaneous, often emotional, and rarely analytical and reflective. As such, it contains prejudice, bias, truth and error, inspiration, and distortions; in short, good and bad reasoning, all mixed together. Second order thinking is essentially first order thinking "raised to the level of conscious realization," that is, analyzed, assessed, and thereby reconstructed.[3]

---

[1] We obtained the basis for much of the information in this section from handouts, discussions, and workshops at the 12th to 20th International Conferences on Critical Thinking and Educational Reform sponsored by Sonoma State University's Center for Critical Thinking and Moral Critique.

[2] Barzun, J. 2000. *From Dawn to Decadence: 1500 to the Present* (New York: HarperCollins), p. 54.

[3] Paul, R., & L. Elder. 2000. *Critical thinking—Tools for taking charge of your learning and your life* (Upper Saddle River, NJ: Prentice Hall).

Many scientists equate critical thinking with the application of the scientific method, but we think critical thinking is a far broader and more complex process. Critical thinking involves developing skills that enable you to dissect an issue (*analyze*) and put it together (*synthesize*) so that interrelationships become apparent. It involves searching for *assumptions,* the basic ideas and concepts that guide our thoughts. Critical thinking also encourages an appreciation for our own and for others' *points of view,* which is important when approaching complex environmental issues.

Too often, analyzing complex issues leads some to a belief that everyone is "entitled" to an opinion that should be respected. We do not necessarily concur. However, problem solving demands a willingness to listen for *content* to what others are saying. *Talking is easy, but listening is not.* Developing critical thinking skills is not like learning to ride a bicycle. All of us must learn to use a set of intellectual standards as an "inner voice" by which we constantly test and hone our reasoning skills. But the standards must be set in an appropriate framework in order for true critical assessment to take place.

The following paragraphs describe the intellectual standards you should apply when assessing the quality of your reasoning. This is the basis for critical thinking, which in turn is the approach that we try to apply throughout this book.

## Intellectual Standards: The Criteria of Solid Reasoning

*Clarity.* This is the most important standard of critical thinking. If a statement is not clear, its accuracy or relevance cannot be assessed. For example, consider the following two questions:

1. What can we do about global climate change?
2. What can citizens, regulators, and policy makers do to make sure that greenhouse gas emissions from industry, transportation, and power generation do not cause irreversible ecological damage or harm human health?

*Accuracy.* Is the statement true? How can we find out? A statement can be *clear* but not *accurate.*

*Precision.* Can we have more details? Can you be more specific? A statement can be clear and accurate, but not precise. For example, we could say that "There were more sport utility vehicles (SUVs) in the United States in 2012 than in 2001." That statement is clear, it is accurate, but how many more SUVs are there? 1? 1,000? 1,000,000? (Note that there is a difference between the way many scientists use the word *precision* and the more general way it is used here.)

*Relevance.* How is the statement or evidence related to the issue we are discussing? A statement can be clear, accurate, and precise, but not relevant. Here's an example: If we are given the responsibility to eliminate the harmful environmental impact of pollutants from coal-burning power plants, and we invite public comment on our proposals, someone might say, "Electricity from coal-burning power plants accounts for 100,000 jobs in this state alone." That statement may be clear, accurate, and precise, but it is not relevant to the specific issue of removing pollution (although in other contexts it may in fact be very relevant).

Here is another example: Parks in Arlington, Virginia, have signs at all entrances that read "Dogs must be on leash and under control at all times" and include the relevant County Code citation. Nevertheless, dog owners typically ignore the signs. Here are some of their reasons: "The neighbors don't complain," "There is no dog exercise area near my home,"

"My dog is well-behaved and doesn't need a leash," and "The police don't mind if our dogs play here." Using critical thinking, assess the relevance of the dog owners' responses.

*Breadth.* Is there another point of view or line of evidence that could provide us with some additional insight? Is there another way to look at this question? For example, you will assess the issue of turf grass proliferation in Issue 14. In an article one of us (RLM) wrote for the *Washington Post,* he suggested that since turf-care devices such as mowers, trimmers, blowers, and the like are significant sources of air pollution, and since their use is proliferating, it might be easier to address the problem they pose not by banning or overregulating these devices, but by reducing the area of turf that must be maintained.

*Depth.* How does a proposed solution address the real complexities of an issue? Is the solution realistic or superficial? This question is one of the most difficult to tackle, because here is where reasoning, "instinct," and moral values may interact. The points of view of all who take part in the debate must be carefully considered. For example, politicians have suggested "Just don't do it" as a solution to the problem of teenage drug use, including smoking. Is that a realistic solution to the problem, or is it a superficial approach? How would you defend your answer? Is your defense grounded in critical thinking?

*Logic.* How does one's conclusion follow from the evidence? Does the conclusion really make sense? Why or why not? When a series of statements or thoughts are mutually reinforcing and make sense together, and when they exhibit the intellectual standards described above, we say they are logical. When the combination does not make sense, is internally contradictory, or not mutually reinforcing, it is not "logical." Logic in an argument is to the trained mind a bit like the apocryphal definition of obscenity: "You know it when you see it" (see the following section on "Logical Fallacies and Critical Thinking").

## Applying Intellectual Standards in a Critical Thinking Framework

The intellectual standards described above are essential to critical evaluation of environmental issues, but there are more factors to be considered. The following criteria constitute the framework in which these standards should be applied.

- **Point of view.** What viewpoint does each contributor bring to the debate? Is it likely that someone who has a job in a weapons plant would have the same view on military spending as someone who doesn't? Would a tobacco company executive be likely to have the same opinion on restricting smoking as someone who lost a relative to lung cancer? Think of other examples, but note that *identifying a point of view does not mean that the opinion should automatically be accepted or discounted.* We should strive to identify our own point of view and the bases for this, we should seek other viewpoints and evaluate their relevance, and we should strive to be fair-minded in our assessment. Few people are won over by having their opinions ridiculed. It is important to note here that our points of view are often informed by our *assumptions,* which we will address below.
- **Evidence.** Scientific problem solving is, or should be, based on evidence and information (sometimes called *data,* but we prefer to apply this term solely for numbers used in calculations). Our conclusions or claims must be based on sufficient relevant evidence, the information must be laid out clearly, the evidence against our position must be evaluated, and we must be open to new evidence that challenges our conclusions.
- **Purpose.** All thinking to solve problems has a purpose. It is important to have a clear understanding of that purpose and to ensure that all participants are on the "same wavelength." Since it is easy to wander off the subject, it is advisable to periodically check to make sure the discussion is still on target. For example, stu-

dents working on a research or term project or employees tackling a work-related problem occasionally stray into subjects that are irrelevant and unrelated, although they may be interesting or even seductive. It is vitally important, therefore, that the issue being addressed must be defined and understood as precisely as possible.

■ **Assumptions.** Here is an excerpt from a January 2001 report prepared by the U.S. Energy Information Administration: "With a growing economy, U.S. energy demand is projected to increase 32 percent from 1999 to 2020, reaching 127 quadrillion BTU, *assuming* no changes in Federal laws and regulations." This statement is clear, precise, and contains assumptions.

All reasoning and problem solving depends on *assumptions*, which are *statements accepted as true without proof*. When we assume something, we presuppose it or take it for granted. For example, students show up in class because they assume their professor/teacher will be there. "Never assume" is an old adage. However, it is more reasonable to be aware of and take care in our assumptions and always be ready to examine and evaluate them. They often need to be revised in the light of new evidence.

Now, before we analyze our own assumptions, let's summarize some characteristics of sound reasoning.

Skilled reasoners

■ Understand key concepts (such as externalities) and ideas.
■ Can explain key words and phrases (such as global warming).
■ Can explain scientific terms (such as bioaccumulation).
■ Continually exercise their thinking skills.
■ Can recognize irrelevant topics and can explain why they are irrelevant.
■ Come to well-reasoned conclusions and solutions.

Additionally, the effective reasoner continually assesses and reassesses the quality of his or her thinking in light of new evidence. Finally, the effective reasoner must be able to communicate effectively with others.

## LOGICAL FALLACIES AND CRITICAL THINKING

After you leave this course, much of your information concerning environmental issues will come from the media—television, magazines, radio talk shows, and newspapers. These sources often exhibit evidence of poor reasoning—that is, logical fallacies. Learn to identify them to ensure you are getting the best information possible. Here, in no particular order, are some of the more common examples.

■ The fallacy of *composition*: assuming what is good for an individual is good for a group. An example is standing at sports events—an advantage to one person but not when everybody does it.
■ The fallacy of *starting with the answer*: including your conclusion in your assumptions. For example, "America runs on high levels of energy consumption. Without energy, we wouldn't have the American lifestyle. We can't have the American lifestyle without energy. Thus, America can't afford to conserve energy." Here, the arguer is simply defining his way out of the problem. By his reasoning, we would have to continue to increase energy consumption forever, an obvious impossibility.
■ The fallacy of *hasty generalization*: "Senegal and Mali have very low levels of energy consumption. They are very poor countries. Low levels of energy consumption lead to poverty." How is poverty measured? Are there any "wealthy" countries that have relatively low levels of energy consumption? Are there any relatively poor countries that have high levels of energy consumption?

- The fallacy of *false choice*: stating an issue as a simplistic "either-or" choice when there are other more logical possibilities. "Those who don't support fossil fuel use want to go back to living in caves."
- Fallacy of an *appeal to deference*: asserting an argument because someone famous supports it.
- Fallacy of *ad hominem* (literally, "at the person") argument: attacking a person or a person's motives without discussing the merits of his or her position.
- The fallacy of *repetition*—the basis of most advertising: repeating a statement over and over without offering any evidence. "Population growth is good. People contribute to society. We need population growth to survive."
- The fallacy of *appealing to tradition*: "Coal built this country. Reducing coal use would threaten our society."
- The fallacy of *appealing to pity*: "Three million people work in the mining industry. Regardless of its impact, we have to support them."
- The fallacy of an *appeal to popularity*: "Seventy-five percent of Americans support this position." Perhaps the poll asked the wrong question; perhaps the respondents didn't have enough information to properly respond, etc.
- The fallacy of *confusing coincidence with causality*: "After passage of the Endangered Species Act (ESA), jobs in sawmills fell 70%. Therefore the ESA was bad for the economy." Were there other possible explanations for the drop in jobs?
- The fallacy of the *rigid rule*: "Hard-working people are good for the economy. Immigrants are hard-working people. Therefore, the more immigrants we have, the better for our economy." Or, "Large numbers of immigrants commit crimes. Crimes are bad for the economy. Immigration should be ended."
- The fallacy of *irrelevant conclusion*: using unrelated evidence or premises to support a conclusion. "Development raises the value of land. Development provides jobs. Developed land pays more taxes than undeveloped land. Therefore flood plain land should be developed."

### Assumptions about Environmental Issues

We cannot stress too heavily the power of assumptions in guiding our reasoning. To give you an example, in the following passages we would like you to respond to the following real-world issues:

First, identify your assumptions in defining government's role in protecting the environment.

Second, identify your assumptions as to the proper *level* of government that may act.

Third, determine your position on the "precautionary principle."

Fourth, identify your assumptions concerning the extent to which individuals or institutions in a society may impose costs upon others in that society with or without their knowledge or consent.

## THE ROLE OF GOVERNMENT

Respond to the following quote, taken from Thomas Jefferson's First Inaugural Address, delivered on March 4, 1801. Most politicians and most Americans probably consider themselves to have "Jeffersonian" principles.[4] Here is what Jefferson said:

> What more is necessary to make us a happy and prosperous people? Still one thing more, fellow citizens—a wise and frugal Government, <u>which shall restrain men from injuring one another</u>,

---
4 See for example www.lewrockwell.com/vance/vance17.html.

shall leave them otherwise free to regulate their own pursuits of industry and improvements, and shall not take from the mouth of labor the bread it has earned.[5]

**Question 11:** In a clear sentence or two, explain what you think Jefferson meant by the phrase we underlined.

**Question 12:** Do you think he was referring solely to thugs who physically brutalize their fellow citizens? Explain.

**Question 13:** Could he logically have been referring also to citizens who sought to poison others? In other words, is restraining poisoners a legitimate role of government? Explain.

**Question 14:** Now, what if a citizen or an organization dumps a toxin into water or air that all citizens depend on, or if a citizen or an organization fills in a wetland that performed valuable ecological functions upon which local residents depend? May government under Jefferson's principle restrain that person or organization?

Your answer to these questions will define your assumptions as to the proper role of government.

## The Proper Level of Government That May Act

Next, we will ask you to evaluate your assumptions about the *level* of government that may properly intervene in environmental issues.

One of the major discoveries of the past two decades has been the extent to which much pollution is *transboundary* in nature. For example, as much as one-third of the NOx (oxides of nitrogen) air pollution affecting the Washington, DC, metropolitan area annually comes from as far away as the Midwest and southern Canada; much of the pollution

---

[5] See http://avalon.law.yale.edu/subject_menus/inaug.asp.

that degrades air quality over Grand Canyon National Park comes from southern California, several hundred kilometers to the west; some of Oregon's air pollution comes from coal-fired electric utilities in China; and so forth.

**Question 15:** Therefore, is it appropriate that local government primarily or solely bear the responsibility for protecting its own environment? May the states and federal government have a legitimate role based on the transboundary nature of pollution? Explain and justify your answer, using additional paper if necessary.

Your answer will help evaluate your assumptions about the extent to which state and federal government agencies have responsibilities to intervene to protect local environments. Remember to reassess your assumptions in the light of new evidence.

## The Precautionary Principle

Scientists generally define the *precautionary principle* as follows:

> Action should be taken to prevent damage to the environment even in cases where there is no absolute proof of a causal link between emissions or activity and detrimental environmental effect. Embedded in this is the notion that there should be a reversal of the "burden of proof" whereby the onus is now on the operator to prove that his action will not cause harm rather than on the environment to prove that harm (is occurring or) will occur. [6]

Another way to express this principle is "better safe than sorry." Many products of science and technology are brought to the marketplace without adequate knowledge of possible long-term effects on human health and the global environment. Some examples are the uses of freon, mercury, and organochlorines, which you will investigate later.

In most industrialized nations, the so-called "burden of proof" falls not on the producers of goods but rather on those who allege that they have suffered harm. This is the basis of our tort system of civil law. As a result of the proliferation of new products, government agencies like the Food and Drug Administration, the Environmental Protection Agency (EPA), and the Federal Trade Commission, to name but a few, are sometimes unable to keep pace. For example, as of 2010, California had registered more than 900 pesticide active ingredients that were used in approximately 12,000 pesticide products. One of six boards and departments within Cal/EPA, the Department of Pesticide Regulation, regulates the sale and use of pesticides to protect human health and the environment.

Although individuals have recourse to law if they believe they have been injured, future generations, wildlife, and ecosystems have no such means of redress. Adherence to the precautionary principle could in the view of many facilitate democratic oversight.

Similarly, a serious threat like global warming or the proliferation and buildup of organochlorines under the precautionary principle would trigger action to address the threat, even if the "science" is not yet conclusive but is supported by the preponderance of available evidence.[7]

---

[6] Glegg, G., & P. Johnston. 1994. The Policy Implications of Effluent Complexity. In *Proceedings of the Second International Conference on Environmental Pollution* (London: European Centre for Pollution Research), Vol. 1, p. 126.

[7] Shabekoff, P. 2000. *Earth Rising: American Environmentalism in the 21st Century* (Washington, DC: Island Press).

**Question 16:** Discuss your opinion on the precautionary principle. Should those who wish to introduce a new chemical, a new industrial process, a land-use change, and so on, have to demonstrate that their change will not harm the environment before proceeding? Explain and defend your answer, using additional paper if necessary.

## The Question of Externalities

Economists define *externalities* as any *cost* of production not included in the *price* of the good. An example would be environmental pollution or health costs resulting from burning diesel fuel not included in the price of the fuel.[8] Another example would be cleanup costs paid by governments resulting from animal waste degradation of water bodies from large-scale meat processing operations. In this example, the price of chicken or pork at your local supermarket is lower than it would be if all environmental cleanup costs were included in the price of the meat.[9]

**Question 17:** Consult an economics textbook or do a search on the Internet using the term externalities. State whether you conclude that externalities should be included in the costs of goods, or whether and in what circumstances some costs can be left for others to pay. Justify your answer using those principles of critical thinking outlined previously.

## Assumptions about Corporations

Here is another quote from Thomas Jefferson, on the impact of those new organizations called corporations. Read Jefferson's words and then respond to the following question.

> I hope we shall take warning from, and example of, England, and crush in its birth the aristocracy of our moneyed corporations, which dare already to challenge our Government to trial, and bid defiance to the laws of our country.

**Question 18:** Do you share or reject Jefferson's opinions concerning corporations? Cite evidence or provide support to your conclusion. It might help you to prepare a list of positive and negative contributions corporations make to our economy. Do you feel corporations have too much power in contemporary life? Why or why not? Use additional paper if necessary.

If you are interested in corporate power, research the 1886 U.S. Supreme Court ruling: *Santa Clara County v. Southern Pacific*. The Court ruled that Southern Pacific was a "natural person" entitled to the protections of the U.S. Constitution's Bill of Rights and Fourteenth Amendment.

---

[8] For background on diesel, see California Air Resources Board, www.arb.ca.gov/homepage.htm.
[9] See for example Environmental Defense www.environmentaldefense.org.

Then review "Citizens United vs Federal Election Commission" a major 2010 decision that held that, under the 1st Amendment to the U.S. Constitution, the federal government cannot limit political contributions from corporations or labor unions. The court did uphold the existing prohibition on donations by these groups to individual political candidates.

After researching these cases, answer these questions:

**Question 19:** Discuss whether you believe the Court acted correctly in deciding that a corporation was a person.

**Question 20:** Is the 1886 ruling relevant to the twenty-first century? Why or why not?

**Question 21:** Do you agree with the "Citizens United" decision (a 5–4 decision by the way)? Why or why not?

After having thoughtfully responded to the above scenarios, you should now have a better awareness of the assumptions that you bring to the analysis of environmental issues that you are about to undertake.

## SUMMARY

To summarize, the intellectual standards by which critical thinking is carried out are clarity, accuracy, precision, relevance, breadth, depth, and logic. These standards are applied in a framework delineated by points of view, assumptions, evidence or information, and purpose. We encourage you to return to this section whenever you need to refresh and polish your critical thinking skills.

# PART ONE
# Principles of Sustainability

On September 22, 2009 President Barack Obama addressed the United Nations. He began his speech: "The threat from climate change is serious, it is urgent and it is growing. Our generation's response to this challenge will be judged by history, for if we fail to meet it—boldly, swiftly and together—we risk consigning future generations to an irreversible catastrophe."[1]

Seventeen years previous, in 1992, 1700 scientists, including the majority of living Nobel laureates in the sciences, issued the World Scientists' Warning to Humanity:

> Human beings and the natural world are on a collision course. Human activities inflict harsh and often irreversible damage on the environment and on critical resources. If not checked, many of our current practices put at serious risk the future that we wish for human society and the plant and animal kingdoms, and may so alter the living world that it will be unable to sustain life in the manner that we know. Fundamental changes are urgent if we are to avoid the collision our present course will bring about.[2]

The solution to these environmental problems is sustainability.

**Question P1-1:** Why do you think so little was done to address climate change between 1992 and 2009?

## WHAT IS *SUSTAINABILITY*?

Sustainability, or sustainable development, was defined in 1987 by the World Commission on Environment and Development as development that meets the needs of the present without compromising the ability of future generations to meet their own needs.[3] Paul Hawken, author of *Natural Capital*, and *The Ecology of Commerce*, defined sustainability more practically: "Leave the world better than you found it, take no more than you need, try not to harm life or the environment, make amends if you do."[4]

Sustainability transcends and supersedes environmentalism. It involves a transformation from a wasteful linear model of resource use in which natural resources are extracted, converted to goods, then trashed or landfilled, to a cyclical model built around efficiency, waste reduction, reuse, and recycling. Moreover, environmentalism focuses primarily on the natural, nonhuman world and the effects humans have on it, whereas sustainability adds social and economic justice components. Thus, sustainability is based on what is known as the *triple bottom line*: planet, people, and prosperity.

---

[1] http://cop15.state.gov.

[2] http://www.ucsusa.org/about/1992-world-scientists.html.

[3] World Commission on Environment and Development (Gro Harlem Brundtland, Chair). 1987. *Our Common Future* (New York: Oxford University Press).

[4] Hawken, P. 1994. *The Ecology of Commerce* (New York: HarperCollins).

# MEASURING SUSTAINABILITY

Measuring sustainability is not straightforward. In 2005 the Environmental Performance Measurement Project at Yale University issued an Environmental Sustainability Index (ESI), ranking nations on the extent to which their societies approached sustainability. This Index was updated in 2012.[5] The countries ranked as "strongest performers" were in order: Switzerland, Latvia, Norway, Luxembourg, Costa Rica, France, Austria, Italy, United Kingdom, and Sweden. The U.S. ranked 49th. Iraq ranked last, at 132. In most cases, however, insufficient data exist to accurately determine each nation's ESI. Their rankings are, therefore, only crude comparative measurements of societal sustainability.

The Sustainable Communities Network sets general targets for a Sustainable Planet Earth. They are[6]

Creating Community
Growing a Sustainable Economy
Protecting Natural Resources
Governing Sustainably
Living Sustainably

In our view achieving sustainability requires addressing agriculture, fisheries, forests and wood products, water supplies, energy, biodiversity, climate change, manufacturing and industry, and justice and equity. Each of these is addressed as individual Issues, or within Issues.

## Agriculture

A sustainable society must preserve agricultural land, practice sustainable agriculture, and produce substantial food supplies locally. Agriculture should be based ultimately on organic methods. In the United States and European Union (EU), the consumption of organic food has been increasing at 15 to 20 percent annually for more than a decade. Progress toward sustainable agriculture has been slowed by agricultural subsidies, which distort markets and often harm the environment. In 2012 *The Economist* estimated agricultural subsidies from the EU, the U.S., Japan, China, Russia, and Brazil alone at $360 billion.[7]

Eventually, the impact of the growing global demand for meat must be addressed. Contemporary industrial-style meat production imposes significant environmental costs. For example, cattle-raising poses one of the greatest threats to the survival of tropical rainforests. Moreover, per capita yields from global fisheries are declining, and expansion of "aquaculture" is often at the expense of coastal ecosystems and wild fish stocks (see "Fisheries").

## Fisheries

Many fish species ranging from sardines to bluefin tuna are in dangerous decline. Thriving and diverse aquatic wildlife are necessary for healthy marine and freshwater ecosystems. It is therefore critical that communities dependent on fisheries and aquatic ecosystems use these resources responsibly. One successful method is creating marine protected zones that are large enough to be self-sustaining, thus protecting species diversity from human interference. But protected zones alone may not be enough: Many large fish and marine

---

[5] For data on which these rankings were based, go to http://epi.yale.edu/epi2012/methodology.
[6] www.sustainable.org.
[7] See http://www.economist.com/node/21530130.

March 6, 2006 (Terra ASTER)

**FIGURE 1** Satellite view of shrimp farm ponds in Ecuador. Darker rectangles represent shrimp farm ponds that have encroached upon natural landscapes, including mangrove ecosystems. According to the UN Food and Agricultural Organization, in 1999 Ecuador was the fourth largest producer of shrimp in the world, due almost exclusively to conversion of wetlands to shrimp farms. (Courtesy of NASA.)

mammals have dangerously high levels of toxic artificial chemicals such as organochlorines in their tissue.[8]

Protecting aquatic wildlife could be aided through sustainable aquaculture. For example, growing herbivorous fish like carp and tilapia puts less strain on resources compared to growing carnivorous species like salmon, which usually must be fed with feed made from wild fish. Shrimp farming puts great stress on coastal ecosystems, since mangrove communities are often cleared to make room for shrimp ponds. Figure 1 shows rectangular shrimp ponds on the coast of Ecuador, taken by National Aeronautics and Space Administration (NASA). Many oceanic species have been decimated by industrial-style fishing practices as well as by massive national subsidies for fishing fleets.

Ironically, the destruction of ocean fisheries coincides with an increased demand for fish resulting from its recognition as a health food.

## Forests and Wood Products

Trees have economic value as a raw material, yet the environmental services provided by forests far transcend the economic value of trees. In addition, trees are important for urban communities and essential for the moderation of global climate. Mature trees maintain desirable microclimates and shelter wildlife. Tropical rainforests actually generate their own precipitation. Figures 2a and 2b show Africa's Mt. Kilimanjaro, one of the planet's most awesome volcanoes. The "snows of Kilimanjaro" immortalized by Ernest Hemingway are rapidly vanishing. Figure 2b shows the volcano in February of 2000, with snows nearly gone. In addition to global climate change, one of the reasons snowfields are disappearing from Mt. Kilimanjaro is the destruction of forests around and on the giant vol-

---

[8] See for example http://www.ukmarinesac.org.uk/activities/water-quality/wq8_42.htm.

A

B

**FIGURE 2** Mt. Kilimanjaro, whose snowcap is suffering from the dual insults of global warming and deforestation. A: Mt. Kilimanjaro's full snowcap in the late 20th century. (Courtesy of Anup Shah/naturepl. com.) B: A view of the volcano in February of 2000 showing the loss of snow. (Courtesy of NASA.)

cano—forests that transpire sufficient moisture to fuel precipitation, which sustained the snowfields.

In the 13 states of the U.S. Forest Service's Southern Region, forests cover 214 million acres, which amounts to 29% of total forest cover in the United States. Yet, virtually none of the forest in the Region remains pristine, since 99% of southern forests have been cut in the last 400 years. The majority of southern forests are *commercial* forests, typically meaning rows of loblolly pines or other fast-growing trees scattered among recently harvested clearcuts. Known as *tree plantations,* these areas may superficially resemble forests, but they are *monocultures* (composed of a single kind of tree), their biodiversity is not as high as that of undisturbed, natural forests, and they do not provide the same levels of ecosystem services (see below) that natural forests provide and on which humans depend.

Forests protect water supplies and provide habitats, enhancing species diversity. They store carbon, mediating climate change. Healthy forests are essential to sustainable societies. In North America, this means forests with diversity levels approaching those encountered when Europeans first colonized the hemisphere.

## Water Supplies

High-quality water supplies are essential both for human use and to maintain the health of local ecosystems. Protection of global water supplies and aquatic ecosystems in the face of growing human populations will be one of humanity's greatest challenges. Reducing waste of water in irrigated agriculture is one of the easiest ways to increase water supplies. Subsidies for irrigation, however, often encourage waste over conservation.

In the United States, Western water laws and doctrines often require those with water rights to use them or lose them. However, change is possible and is indeed underway. In California, agreement has tentatively been reached between irrigators in the Imperial Valley and other stakeholders to share water from the Colorado River and to ultimately reduce water use. However, Mexican farmers, who depend on irrigation water leaking from unlined canals across the border in the Imperial Valley, are suing to prevent sealing of irrigation canals.

## Energy

Sustainable societies cannot be built on nonrenewable energy resources. Humans use almost unimaginable amounts of energy and generate vast amounts of pollution. Reducing pollution from fossil fuels requires, at the very least, that laws be strictly enforced. For example, according to the Vermont Journal of Environmental Law, "Chinese environmental laws and regulations are abundant, but suffer from a lack of proper adherence and enforcement," resulting in the needless production of pollutants like SOx and NOx (oxides of sulfur and nitrogen), particulates, and heavy metals like mercury and uranium.[9]

Pollution imposes significant, measurable costs on human and ecosystem health. Energy conservation and the use of renewable fuels provide cost-effective and sustainable alternatives that generate little air and water pollution. Subsidizing the production of coal and oil-based nonrenewable energy makes little sense in a world threatened with rapid climate change and accelerating species loss, in part due to the burning of fossil fuels.

Here, too, change is coming. Wind energy is the fastest growing energy source in Europe and North America, supported by government subsidies which partly offset subsidies for fossil fuels. Energy companies like NextEra Energy and General Electric are major producers and developers of renewable solar and wind energy. And although biofuels cannot yet replace fossil fuels in transportation, producing transport fuels from agricultural waste, and even waste cooking oil, can make a dent in expensive imports and reduce air pollution at the same time.

## Biodiversity

Habitat loss is the greatest threat to biodiversity on the planet.[10] Increased food production, including meat, for a richer, more populous Earth is a major cause of habitat loss. Over half of global forests have already disappeared, and they are being removed at a rate, 17 million hectares a year, ten times greater than maximum restoration rates.

Our very survival as a species could ultimately rely on maintaining the integrity of ecosystems we barely understand. An ecosystem is a geographic area including all the living organisms (people, plants, animals, and microbes); their physical surroundings (such as soil, water, and air); and the natural cycles that sustain them (such as the hydrologic cycle). All of these elements are interconnected. Altering any one component affects the others in that ecosystem. Ecosystems can be small, like a single stand of trees, or large, like an entire watershed.

Biodiversity is particularly critical for sustainability because of the specialized and often little understood roles each species plays in maintaining the dynamic state of ecological balance. Moreover, surprisingly little is known about key ecosystems like soils and the deep ocean.

Esthetics and ethics must also play a part since humans can survive, after a fashion, on an Earth with drastically reduced species diversity. The question then becomes, do we wish to make a decision for future generations to eradicate species and ecosystems, without the input of our descendants? That our ancestors did so in ignorance is no excuse for our perpetuating such behavior.

## Climate Change

While climate change is a well-documented fact of planetary history—the Earth has gone through several megacycles (100,000,000- to 1-billion-year cycles) of "greenhouse" and "icehouse" conditions—it is the speed with which human-induced climate change

---

[9] VT Journal of Environmental Law, www.vjel.org/journal/pdf/VJEL10058.pdf.
[10] See for example The International Year of Biodiversity, http://www.cbd.int/2010/biodiversity/.

is occurring that is unprecedented. Too-rapid change overwhelms the ability of natural ecosystems to adapt, which is exacerbated by the fragmentation of ecosystems by human activity. For example, the rapid acidification of the oceans will likely help exterminate coral reefs by mid-century unless checked, with serious implications for the entire ocean ecosystem. The impacts of climate change are imperfectly understood, but on the whole will certainly test the creativity and capabilities of a human species that "subdued" a seemingly limitless Earth. Sustainable societies may well be essential to address the impacts of climate change.

### Manufacturing and Industry

The Industrial Revolution generated wealth (for some) beyond humanity's dreams, but also generated waste in unprecedented quantities, far beyond the capacity of natural systems to process. Pollution is one form of waste. Wasteless production must become the norm in human activity. And progress is being made: The EU has set a goal of no more waste going to landfills by 2025.

In nature, waste does not exist. Waste eventually becomes something else's food. In human societies, waste is everywhere. Waste indicates inefficiency. Waste can also harm human health and degrade the environment. Many businesses have found that waste reduction and even elimination can enhance profitability. For example, Waste Management Corporation is using landfill gas (methane) to power a sizable proportion of its 22,000 waste collection vehicles,[11] and the Clorox Corp. is phasing out the use of dangerous chemicals in chlorine production. Much progress has been made in this arena—humans have agreed to phase out or eliminate the most harmful kinds of Persistent Organic Pollutants (POPs; see Issue 17). Many destructive chlorofluorocarbons (CFCs) are being phased out globally under the Montreal Protocol, even though the United States has relentlessly sought exemptions for agricultural users of methyl bromide. However, the growth of human populations and the universal association between increasing wealth and increasing waste poses critical problems for a world, five-sixths of whose population is trying to develop along Western-style free-market lines.

### Justice and Equity

The pursuit of justice and equal opportunity are key ingredients in a sustainable civilized society. Examples of injustice are lack of adequate housing, health care, lack of access to education, poor sanitation, an inadequate supply of pure water, exposure to environmental toxins, and environmental degradation related to industrial pollution. Rich societies ignore these issues at their peril.

## SUSTAINABLE CONSUMPTION: AN OXYMORON?

We are in the twilight of the era of the "myth of unlimited resources."[12] Humans contribute to local and global sustainability by adopting responsible patterns of buying, consumption, and reproduction, thereby consuming minimal energy and fewer resources. For example, conventional methods of construction do not lend themselves to minimizing energy consumption and waste. Construction and Demolition (C&D) Waste is accordingly a major, and often unnecessary, component of Municipal Solid Waste. Responsible consumption is based on education not coercion in a democratic society. Unfortunately, industrial and postindustrial societies are philosophically devoted to ever-increasing consumption, in turn driving ever-increasing production—the "growth" concept. Detoxifying society

---

[11] See for example http://www.wm.com/sustainability/renewable-energy.jsp.

[12] See for example http://www.jayhanson.us/page130.htm.

from the "unlimited consumption" myth (perhaps by favoring service consumption rather than material consumption) may be one of our greatest challenges. Sustainable societies are probably incompatible with ever-increasing numbers of "self-storage" facilities, for example.

## HEALTH AND NUTRITION

Poor individual physical and mental health imposes significant costs upon society, in the form of health-care expenses, crime, and lost productivity, for example. With all the environmental toxins loosed on the planet by human activity, the greatest killers in wealthy societies remain diseases related to smoking, alcohol, drugs, and obesity. Estimates of the total health and productivity costs of cigarette smoking run as high as $193 billion per year. And the Center for Disease Control and Prevention (CDCP) reports that medical spending on obese persons is $1400 a year more than otherwise. Twenty-seven percent of the U.S. population is now obese and the proportion is growing.[13] While cigarette smoking is declining in America, companies aggressively export the habit to developing countries, a practice that is counter to notions of fairness and equity.

## SUSTAINABLE POPULATION

Human numbers must eventually become stabilized, since it is physically impossible for growth to continue forever. The only questions are at what level will growth end and whether growth will end as a result of human actions or by natural processes like famine, disease, and war. Aging societies are typical of developed nations (e.g., the United States). Populations dominated by the young are typical of developing ones (Vietnam, for example). Large numbers of young people provide great promise for societies, but also impose great costs, especially in countries like Spain and Egypt in which up to half the young population of working age (roughly from 16 to 40) is unemployed. Stable, aging societies face challenges of paying for retirement benefits, if such benefits are based on government programs like Social Security, which generally pay to retirees far more in benefits than they paid in.

The readers of this book are mainly young, and it is they who will solve, or not solve, these challenges. The next century should prove to be one of the most interesting, and potentially rewarding, centuries in the entire span of human history.

## DEVELOPMENT

A central question in *ecological economics* is to what extent *development* (leading, as it is supposed, to higher per capita income), global *trade*, and a country's *environmental quality* are related.

### What Is Development?

Here we use development to refer to a complex set of changes which convert the economy of a society based on subsistence agriculture to one in which most of the employed inhabitants work in manufacturing or services. Early stages of development typically depend more on the exploitation of "natural resources," and later stages on "human capital," that is, the creativity of the human mind.

Development has historically involved significant land-use changes, including

- *deforestation* for fuel wood and to provide land for intensive farming
- *urbanization*, and

---

[13] See for example http://www.cdc.gov/obesity/data/adult.html.

   ■   *large-scale mining* for fossil fuels and metals.

   Development has also produced unprecedented quantities of environmental pollution
and waste. These include exhaust gases from the burning of fossil fuels and toxic factory emissions polluting waterways. As some consistently point out, such urban pollutants
as horse urine and droppings have declined, but pet waste now constitutes a significant
source of urban pollution. For example, there are at least 500,000 dogs in New Jersey (and
8.8 million people) according to that state's Department of Health.

   Concentrating humans in cities concentrates waste, both human and otherwise, which
if not properly treated can severely pollute waterways or the ocean in the case of coastal
cities.[14] Urbanization may also facilitate epidemics. The cholera and yellow fever epidemics in nineteenth century England and North America are examples. However, the concentration of waste and pollution in cities can make it easier to deal with.

## Development's Impact on the Environment

Here are two hypotheses, much simplified, that purport to explain such relationships as
may exist between development and the environment.

1.   **Development harms the environment.** Many environmentalists point out that de-
     velopment leads to harmful land-use practices, injurious levels of air emissions,
     subsidies encouraging fossil-fuel use, and water pollution, among other things.
     Moreover, they cite high levels of population growth in many developing countries
     as exacerbating environmental decline, leading to misery, child prostitution and the
     like, and encouraging large-scale emigration.
2.   **Development eventually improves the environment.** Many economists and some
     environmentalists, while acknowledging harmful levels of environmental pollution
     in countries in early stages of development, cite considerable empirical evidence
     that (1) population growth rates decline as development proceeds and (2) rates of
     some forms of environmental pollution decline as per capita income increases, a
     supposed corollary of development as we noted above. Newer forms of technol-
     ogy tend to be less polluting than older forms, but also tend to require high capital
     expenditures.

   Countries with higher per capita incomes tend to have cleaner environments along with
increased consumption of goods and services. Poverty and high rates of population growth
are major causes of environmental degradation, such as deforestation. As nations become
richer and their middle classes expand, so do demands for tougher environmental standards
and regulations. Indeed, the "green" movement and "green consumerism" in the developed
world evolved with the growth of the middle class after World War II.

# ECONOMIC GROWTH AND THE ENVIRONMENT—KUZNETS CURVES

The environmental Kuznets curve, named for Nobel laureate economist Simon S. Kuznets,
plots the *relationship between environmental quality factors* and *per capita income*.[15] The
relationships that have been plotted include income with the following: sulfur dioxide
emissions, suspended particulate matter, carbon monoxide, nitrogen oxides, and airborne
lead. In addition, researchers have developed curves by plotting the following environmental parameters against per capita income: access to safe water, presence of urban sanitation

---

[14]   www.state.nj.us/dep/watershedmgt/pet_waste_fredk.htm.
[15]   For example see *New York Times* http://tierneylab.blogs.nytimes.com/2009/04/20/the-richer-is-greener-
curve/.

(that is, whether or not people are connected to sewage treatment plants), annual deforestation rates, total deforestation, dissolved oxygen in rivers, fecal coliform (a measure of the presence of the toxic bacterium *E. coli*) in rivers, municipal solid waste per capita, and carbon emission per capita.

## Shapes of Kuznets Curves

Environmental Kuznets curves (EKC) generally exhibit one of three shapes. One shape results when an environmental benefit improves continually with increasing per capita income. A second shape of Kuznets curves shows a continuous increase in an environmental problem such as municipal solid waste (MSW) with rising incomes. The Kuznets curve that has received the most attention, and has stimulated the most discussion, has an inverted "U" shape.[16] It has been used to suggest the path air quality will follow as economic development, and presumably per capita income, increases. Figure 3, depicting the relationship between per capita income and sulfur dioxide level, illustrates one such curve. Similar U-shaped curves have been reported for particulates.

**Question P1-2:**   Interpret the relationship between income and SOx emissions.

Economists differ on income levels at which the "U" shape in the graphs begin. They varied between $3,000 and $8,700 for sulfur dioxide and ranged up to $10,300 for suspended particulate matter.

**Question P1-3:**   Is it reasonable to conclude that some environmental impacts of economic development are not serious because they will decline over time? Why or why not?

Most types of environmental degradation can be offset at a cost. Scrubbers on power plants can remove up to 90 percent of SOx, for example, and increased per capita income gives nations the wealth with which to afford the expense. Carbon emissions per capita show a pattern similar to municipal solid waste; that is, carbon emissions increase with per capita *gross domestic product* (GDP), the total value of goods and services per individual.

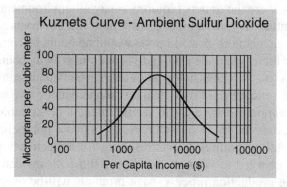

**FIGURE 3** Kuznets curve showing ambient sulfur dioxide concentration and per capita income.

---
[16] Gluskoter, H. 1997. Some environmental effects of increased energy utilization in the twenty-first century. Proceedings of the 17th World Mining Congress.

## Criticisms of Kuznets Curves

Some economists point out that the data are gathered by country, and not globally, and do not take into consideration international trade and the likelihood that wealthier countries are exporting some of their environmental problems to less developed countries. Nor can values for individual countries be extrapolated to the Earth as a whole. However, even when an Environmental Kuznets Curve U-shape relationship is accepted, the *turning point* on the curve, that is, when environmental degradation starts to decline with increasing per capita income, is often found to be very high relative to the per capita GDP of most countries.

In the case of tropical deforestation, researchers found that per capita income levels of most countries in Latin America and Africa were well below the estimated turning point peaks, implying to some that deforestation would eliminate all old-growth forests before those societies could afford to cease the practice.

Such results suggest that the majority of countries have not yet reached levels of per capita income for which environmental improvement is likely to spontaneously occur, unless this improvement is simply dictated by central governments. Worsening global environmental degradation could occur even as the global economy expands and populations grow, and even as some countries make progress cleaning up specific aspects of their environment.

## Global Trade and Environmental Quality

The value of international trade in 2010 approached $15.9 trillion, according to the World Trade Organization.[17] The effect of global trade on environmental quality is controversial. Moving oil by tanker can lead to oil spills. And countries can improve their own environment by "exporting" polluting industries and waste to other nations. For example, during the 1970s and 1980s many metal smelters closed in North America, due to the unwillingness of operators to invest in antipollution technology as required by law. Many of these operations simply relocated outside North America to developing countries, who were willing to tolerate the resulting deterioration of their environment. Similarly, the United States and several western European countries have exported toxic wastes like PCBs to countries like Nigeria, where "disposal" costs were a fraction of those in the home country.

Another example of the adverse impacts of global trade is the introduction of *invasive species* into new environments (see Issue XX). This can occur either "accidentally" as in the ballast water of cargo ships, or on purpose as investments, or for some presumed benefit.

These and other examples too numerous to cite here represent vast subsidies to world trade—in other words, were the traders required to pay all environmental costs associated with their activities rather than dump those costs onto the environment of the receiving country, the volumes and patterns of world trade would no doubt be considerably different.

However, without increased trade leading to economic growth, countries may not have the financial resources to comprehensively address environmental problems.

At the same time, unregulated "free" markets alone cannot guarantee high levels of economic growth accompanied by a clean environment. For example, costs associated with water pollution produced by manufacturers in China are not included in the price of exported goods. Or total air pollution costs resulting from electricity production are not included in the price charged to consumers. Unless some means can link the pollution abatement costs to the production process, such products will be overproduced and will "out-compete" less-polluting alternatives, like electricity from wind.

---

[17] www.wto.org: see for example www.wto.org/english/news_e/sppl_e/sppl236_e.htm.

## Addressing Environmental Impact of Trade and Development

In the case of pollution that crosses national boundaries, addressing the problem usually requires international agreements, and we give two examples below. However, sanctions and punitive tariffs, two oft-proposed remedies, may be counterproductive. As researchers at the Progressive Policy Institute note, "imposing trade sanctions on goods from poor countries with lax environmental standards simply lowers their economic growth and does nothing to counter the poverty that may well be contributing to the environmental problem."[18]Alternatively, international standard-setting organizations could develop voluntary guidelines for labeling, similar to those used to indicate sustainable forestry practices or products of organic agriculture.

Some attempts have been made to address the most egregious problems associated with underregulated trade. The Basel Convention of 1988 began to set controls on the export of toxic wastes from OECD (Organisation for Economic Co-operation and Development) countries to less-developed countries for disposal, for example.

To eliminate the adverse environmental impacts of global trade, many environmental economists, geologists, and ecologists recommend a number of international or multilateral agreements. They include proposals to

- eliminate invasive species from the ballast water of vessels involved in commercial activity in international waterways
- eliminate or tax transboundary air and water pollution
- require "double-hulled" tankers for the shipment of petroleum—already required for shipment of oil in U.S. territorial waters
- promote negotiations to ensure that workers engaged in export industries receive benefits and wages that are at parity with those in the countries to which the goods are shipped
- promote international agreements to regulate such degrading activities as clearcutting and metal smelting so that industries cannot be subsidized by lax environmental standards for moving their polluting activities from one country to another, and
- promote treaties to eliminate, price, or restrict the use of toxic materials, or harmful practices, in agriculture.

## The World Trade Organization (WTO)

The founding charter of the WTO formally addresses the relationship of trade to the environment.[19] It states that signatories to the WTO should "allow for the optimal use of the world's resources in accordance with the objective of sustainable development, seeking to both protect and preserve the environment." WTO rules also allow countries to impose trade regulations "necessary to protect human, animal, or plant life or health" or "relating to the conservation of natural resources." However, measures taken to protect the environment must not discriminate. A country may not be lenient with its domestic producers and at the same time be strict with foreign producers. Nor can member nations discriminate among different trading partners.

The WTO has been controversial since its founding. We will use one example to illustrate the controversy, the issue of sea turtles.[20] Five Asian nations challenged a U.S. law designed to protect sea turtles from certain harmful fishing practices. The law banned the importation of shrimp from countries that did not require the use of turtle-excluder devices (TEDs) by their fishing industry. The WTO dispute panel ruled in favor of the Asian countries, not because it disapproved of United States' attempts to protect sea turtles, but

---

[18]  For example, see http://environmentaleconomics.wordpress.com/.
[19]  See www.wto.org/english/docs_e/docs_e.htm.
[20]  See for example International Environmental Law Project http://law.lclark.edu.

because the panel found the United States had discriminated among members of the WTO, granting preferential treatment to Latin American and Caribbean nations, but not to Asian countries. This decision infuriated environmentalists even though it had arguably nothing to do with the desirability of saving sea turtles.

The WTO cannot dictate national government policy. Sovereign nations choose to become members of the WTO and to play by its rules. So far, more than 140 countries have joined, and others have applied for membership.

## International Agreements to Protect the Environment

Two examples of relatively successful international agreements to protect the global environment are the Basel Accords on Toxic Waste and the Montreal Protocol.

***The Basel Accords on Toxic Waste.*** The Basel Accords on Toxic Waste[21] became effective in 1992. It identified nearly four dozen specific categories of waste to be deemed toxic, and banned export of any of these materials from any signatory nation to any signatory nation that indicated that it did not want to receive such waste. It contained articles defining illegal activity in waste shipment and required all signatory states to eliminate such activity. It stipulated that such waste should be minimized, and disposal or treatment should occur as close to the source as possible.

As of 2010, most nations, including the EU, had ratified the Convention. Three signatories to the Convention had not yet ratified it: Afghanistan, Haiti, and the United States.

***The Montreal Protocol.*** Severe depletion in stratospheric ozone has been measured for years, especially in the Southern Hemisphere. The Montreal Protocol on Substances that Deplete the Ozone Layer was adopted in 1987 as an international treaty to eliminate the production and consumption of ozone-depleting chemicals (ODCs). Four agencies were tasked with implementing the Protocol: the World Bank and three United Nations' agencies—the UN Environment Programme, the UN Development Programme, and the UN Industrial Development Programme.

The Montreal Protocol stipulates that the production and consumption of compounds that deplete ozone in the stratosphere—CFCs, halons, carbon tetrachloride, among others—were to be phased out by 2000, or 2005 in the case of methyl chloroform and methyl bromate. By 2009, the Protocol had achieved universal acceptance, and had achieved a 98% of ozone-depleting chemicals. By mid-2010, developing countries had phased out more than 270,000 tonnes of ozone depleting chemicals.[22]

## Conclusion

World trade has enormous potential to foster the objectives of development, but can also be the source of massive environmental degradation, without multinational and international agreements to "level the playing field." The Basel Convention and Montreal Protocol are two examples of the types of international agreements that could serve as templates for new agreements to address the negative impacts of underregulated international trade.

The debate about the impacts of development continues. Kuznets curves are one, albeit controversial, means of analysis by which data on the consequences of development may be evaluated.

---

[21] www.basel.int.
[22] See for example U.S. EPA http://www.epa.gov/ozone/intpol/.

# Issue 1

## GLOBAL POPULATION GROWTH:
## *IS IT SUSTAINABLE?*

### KEY QUESTIONS

- What is the Earth's population and how fast is it growing?
- Where is population growing fastest?
- How can we project future population?
- How does human population growth threaten the environment?
- What can we do to lessen these threats?
- Who is responsible for solving the problem if there is a problem?
- Is human population growth sustainable?

### INTRODUCTION

In this issue we are going to examine world population growth. Why is human population growth an environmental issue? Although humans exert a substantial positive impact with our enormous ingenuity, we also have a profound physical impact on our local, regional, and global environment. First we, and our buildings and roads, take up space that at one time was forest, wetland, prairie, or hillside. The space an individual occupies may be small, as in cities in Brazil (Figure 1-1), or it may be large, as in new U.S. single-family houses (Figure 1-2), which averaged 2,392 ft$^2$ by 2010.[1]

In addition, since humans must eat, we require land for agriculture, and much of that land is irrigated. In most instances, this land is also fertilized, usually with artificial fertilizers. An additional pollution load is generated by animal waste, especially concentrated by meat production in factory farms (also known as confined or concentrated animal feeding operations, or CAFOs). As a result, fish kills in streams draining agricultural areas and CAFOs are a recurring problem.[2]

People also require transportation. Roads generate polluted runoff, cover permeable land with impervious paving, and take formerly productive land off the tax base. Roads also take an enormous toll on animals (including humans) and other wildlife, divide habitats (which may accelerate species loss), and provide humans easy access to wilderness and forest land, frequently resulting in forest fires and invasion of exotic plant species (by seeds affixing to tires, for example).[3]

---

[1] www.census.gov/const/C25Ann/sftotalmedavgsqft.pdf.

[2] Spills and kills: Manure pollution and America's livestock feedlots: A report by the Clean Water Network, the Izaak Walton League of America, and the Natural Resources Defense Council. 2000 (www.cwn.org/).

[3] Macfarlane, G. 1997. Roads and weeds, partners in crime. *The Road RIPorter*, May/June 1997: 6–7.

**FIGURE 1-1** Crowding in a favela in Salvador, Brazil. (D. Abel)

**FIGURE 1-2** Typical U.S. beach home in the 2000s. (D. Abel)

The waste created by our increasing human numbers is another major environmental problem. In the United States, for example, each person in 2010 produced around 735 kg (1660 lb) of municipal waste per year, nearly half of which was recovered, composted, or burned for electricity.[4] (This does not count sewage, mining, or industrial waste.)

Freer (less-regulated) world trade and a more integrated world economy mean citizens of one country can have a significant impact on the environment in another. A country can import raw materials such as tropical hardwoods or ores, and it can ship toxins to other countries for "disposal."

So, although each additional human being contributes something unique to the planet, each one of us places physical demands as well. Furthermore, people in developed countries

---

[4] http://www.epa.gov/epawaste/index.htm.

like the United States consume disproportionately more resources than residents of poorer countries. Many scientists now conclude that the Earth's ecosystem, the sum of all the planet's smaller ecosystems, may have already reached a point where it can no longer process this demand.

## EXPONENTIAL GROWTH AND ITS IMPACTS

On October 12, 1999, the United Nations "celebrated" its "Day of 6 Billion," the day at which the Earth reached an estimated human population of 6 billion. According to the U.S. Census Bureau, the Earth's population was 5 billion in 1987, 4 billion in 1974, 3 billion in 1959, and 1 billion in about 1825 (Figure 1-3). The Earth's human population was about 7.05 billion in mid-2012, and growing at 1.2 percent per year.[5]

### Calculating Exponential Growth

Before you can analyze the impact of population growth, you need to understand the simple math that describes such growth. When any quantity grows at a fixed rate (or percentage) per year, say 1 percent or 10 percent, as contrasted with growing by a fixed quantity every year, e.g., 80 million people, that growth is said to be *exponential* (Figure 1-3). Exponential growth is calculated using the compound interest formula, also known as the compound growth equation. This is the same formula used to calculate interest in bank accounts. The equation is

$$\text{future value} = \text{present value} \times (e)^{rt}$$

where e equals the constant 2.71828... , r equals the rate of increase, and t is the number of years (or other units) over which the growth is to be measured.

Replacing the words with symbols, the equation reads:

$$N = N_0 \times (e)^{rt}$$

The variable $N_0$ equals the quantity at time zero, i.e., the starting point.

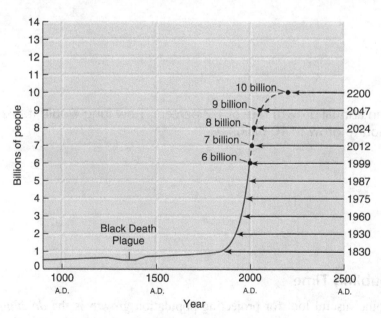

**FIGURE 1-3.** World population growth.

[5] www.prb.org, and see also www.census.gov/ipc/www/worldpop.html.

The compound growth equation allows demographers to project future population size, based on the current population and the population growth rate. Using it is not as intimidating as it may seem. We show you how in "Using Math in Environmental Issues," pages 6–8.

When we use the compound growth equation to project population growth, we make assumptions. (Naturally, we try to determine the reliability of our data.) Primarily, we assume that the population growth rate we are using remains constant over the period we are projecting. Assumptions may not always be accurate. Population estimates are compiled from censuses all over the world. As you might imagine, making accurate estimates in both developing countries and large nations like China and India is difficult. Still, the compound growth equation is a particularly useful tool for examining growth scenarios and planning for the future. Let's practice using it by projecting world population growth.

**Question 1-1:** The mid-year 2012 world population was 7.05 billion and was growing at a rate of 1.2 percent. Project what the world population would be in 2025, 2050, and 2100 at this constant growth rate.

The compound growth equation can also be rearranged. If you know the starting and ending population sizes over a given period, you can calculate the average growth rate over that period using the formula $r = (1/t) \ln(N/N_0)$. Also, you can calculate how long it would take a population of a given size to grow (or decrease) to a different size at a specified growth rate using $t = (1/r) \ln(N/N_0)$.

**Question 1-2:** Given a 1999 world population of 6 billion and a 2012 population of 7.05 billion, calculate the average annual growth rate over that 13-year period.

**Question 1-3:** At an annual growth rate of 1.2 percent, how long would it take a population of 5 billion to grow to 10 billion?

## Doubling Time

Another useful tool for projecting population growth is the *doubling time* formula. For any population that is growing exponentially, the time it takes for the population to double is calculated using $t = 70/r$. In the doubling time formula, in contrast to the other growth formulas, r is entered as the decimal growth rate × 100; thus, a growth rate of 0.07 (that is, 7%) is entered as 7. We explain and demonstrate the doubling time formula in "Using Math in Environmental Issues," pages 6–8.

**Question 1-4:** Use the doubling time formula to estimate how long it would take a population of 5 billion to double given an annual growth rate of 1.2 percent.

**Question 1-5:** According to the U.S. Census Bureau, world population will increase from 6.52 billion in 2006 to 9.30 billion by 2050. Calculate the average annual growth rate for that period.

**Question 1-6:** Use the growth rate you calculated in Question 1-5 to project when the 2012 population will double.

**Question 1-7:** Does a population growth rate of 1.2 percent sound large? Explain then why it might be cause for concern. Be as specific as you can and cite examples of your concerns.

**Question 1-8:** Biologist Garrett Hardin's "Third Law of Human Ecology" states that the total impact of a population on its environment is determined by the absolute population size multiplied by the impact per person. In what ways does your perception of your impact on the planet as a U.S. resident differ from your perception of the impact of a resident of a developing country?

**Question 1-9:** In light of your answer to Question 1-8, what do you think would be more effective in minimizing the impacts of humans on the planet: controlling population growth in developing countries (where most growth is in fact occurring) or curbing consumption of resources in developed countries (where the impact per person is highest)? Cite evidence for your answer. Identify the assumptions you used to answer this question.

**Question 1-10:** Upholding and improving women's rights and interests, including education, reproductive rights, economic security, and health care, have been called the most important actions to control population growth.[6] Explain how this might work. Do you agree or disagree with the statement? Give evidence for your answer.

## POPULATION DENSITY

Using the modification of the compound interest formula introduced earlier, $t = (1/r) \ln(N/N_0)$, we could determine when, at a growth rate of 1.2 percent per year, the world's 2012 population of 7.05 billion would grow to occupy the Earth at a density of one person per square meter (1 person/$m^2$) of dry land. The Earth has $1.31 \times 10^{14}$ $m^2$ of dry land. Thus, this number also represents the population size that would fill the Earth if one person occupied an area of one meter.

**Question 1-11:** Starting with the 2012 world population and using a population growth rate of 1.2 percent, in what year would the Earth reach this impossible density of 1 person/$m^2$ of dry land?

There are places where population densities approaching 1 person/$m^2$ already exist. A two-story building in Delhi, India, was found to house 518 people, a density of 1 person/1.5 $m^2$.[7]

---

[6] World Resources Institute, 1995. *World Resources 1994–95* (New York: Oxford University Press).

[7] World Resources Institute. 1996. *World Resources 1996–97—The Urban Environment* (New York: Oxford University Press).

**Question 1-12:** Calculate the floor area of this building in Delhi. How does this compare to the average floor area of a typical U.S. single-family home (~2,400 ft$^2$ for a new home built in 2010)?

**Question 1-13:** Summarize the main points of this Issue.

**Question 1-14:** Discuss the issue of global population growth from the standpoint of sustainability. To what extent, if any, is a growing human population compatible with a sustainable society? Consider open space and species loss and how they are affected by the increase in human population. Do you feel that these are causes for concern? Can we solve this problem (if it is a problem) by putting animals in zoos or setting aside land for parks or other green spaces? Why or why not?

## For Further Thought

**Question 1-15:** *The Impact of AIDS:* According to the United Nations Population Division, AIDS is having a devastating effect on countries in subSaharan Africa.[8] Life expectancy in the twenty-nine hardest-hit African countries is seven years less than it would have been in the absence of AIDS. In Botswana, the hardest-hit country, one of every four adults is infected with HIV, the virus that causes AIDS. What actions do you think the U.S. and other developed countries should take to lessen the impact of AIDS in Africa?

**Question 1-16:** Do you conclude that the calculations you just performed in this issue transform the discussion from whether population growth will be controlled, to one of how and when? Explain.

**Question 1-17:** Developing countries are experiencing highest rates of population growth. Do richer countries have an obligation to help poorer countries control their growth? If so, what measures would you recommend. Explain.

---

[8] www.census.gov/const/C25Ann/sftotalmedavgsqft.pdf and UN Population Division, The Demographic Impact of HIV-AIDS.

**Question 1-18:**   How do you respond to those whose opinion differs from yours as to whether population growth is a problem? Make a list of significant points in your argument.

**Question 1-19:**   Herman Daly, a University of Maryland ecological economist, believes that population control doesn't deny anyone the "right" to be born, it just asks that they "wait their turn." Do you agree? Discuss and explain your reasoning.

# Issue 2

## CARRYING CAPACITY AND ECOLOGICAL FOOTPRINT

### KEY QUESTIONS

- What factors keep natural populations in balance?
- What does *carrying capacity* mean?
- Can scientists determine the carrying capacity of Earth? Of the United States? Of your hometown?
- What is an ecological footprint?

### BACKGROUND

As he developed his theory of evolution by natural selection, Charles Darwin observed that for any species, more individuals are born than will survive to reproduce. If resources are unlimited and environmental conditions are ideal, the number of offspring reaches a maximum. This state of highest reproductive power, which varies widely among different organisms, is known as a population's *biotic potential.*

However, a number of factors collectively known as *environmental resistance*, normally prevent populations from reaching their full biotic potential and thus growing explosively. Environmental resistance includes disease, predation, drought, temperature extremes, lack of food, and other adverse physical or chemical conditions. Most of these factors are density-dependent: that is, their effects are most pronounced when the population density—the number of individuals in a given area—increases beyond a certain level. Thus, the interplay between biotic potential and density-dependent environmental resistance keeps a population in balance.

### CARRYING CAPACITY

An important ecological concept related to this balancing act is *carrying capacity.* Carrying capacity is the maximum number of individuals of a given species that an area's resources can support in the long term without significantly depleting or degrading those resources. For humans, this definition is expanded to include (1) not degrading our cultural and social environment and (2) not harming our physical environment in ways that would adversely affect future generations.[1]

Determining carrying capacities for most organisms throughout the animal and plant kingdoms is, at least in theory, a reasonably direct calculation. However, when carry-

---

[1] Bouvier, L., & L. Grant. 1994. *How Many Americans?* (San Francisco: Sierra Club Books).

ing capacity is used in a human context, the discussion may become contentious and the resultant calculations subject to dispute. Why is this so?

First, some would argue that the term "human carrying capacity" is meaningless. According to Simon and Kahn, "Because of increases in knowledge, the earth's 'carrying capacity' has been increasing to such an extent that the term has by now no useful meaning."[2]

Second, because air and water pollution do not recognize political boundaries, some scientists define the environment occupied by humans and the resources used by them as a global, rather than a regional or national, entity. For them, only a single estimation of human carrying capacity—the Earth's—is valid and meaningful.

Third, carrying capacity for nonhuman species is calculated by using numerical data and mathematical models. But the assessment of human carrying capacity involves scientists, economists, policy makers, sociologists, theologians, and so forth. These groups often disagree when it comes to defining key carrying capacity concepts.

Finally, other organisms are typically limited by their food supply in a particular region, but humans can import food, if one does not consider the Earth as a single region. Similarly, humans can export waste and pollution by air and water to areas outside their immediate surroundings. Also, by buying raw materials or manufactured goods from outside their region, some can avoid the environmental impact of producing the materials themselves.

## THE IMPACT OF POPULATION

As you can see, carrying capacity is a complicated subject. Nevertheless, the human carrying capacity of the Earth, or a portion thereof, can be estimated, granted that different populations have different impacts based on their technology, consumption, and ethics, as well as the simple number of individuals.

For example, other things being equal, 100 million people with a vegetarian diet would have a different environmental impact compared with 100 million people who consume meat. One way to assess this impact is to compare how vegetarian and meat-eating populations affect water supplies. About 1,000 tons of water are required to produce 1 ton of grain. Globally, 40 percent of all grain goes to meat and poultry production.[3] Thus, high levels of meat consumption put additional stresses on global water supplies.

In terms of energy production and consumption, the impact of industrialized countries varies widely. France generates nearly 80 percent of its electricity by nuclear power and has a significantly different regional energy/environmental impact than China, which generates electricity largely with coal and whose power plants often lack emission controls.

When estimating global or regional human carrying capacity, scientists study environmental changes, as well as the rate that these changes occur. Some useful indicators of environmental change include the rate of topsoil loss (see Issue 19), the rate of species loss, the rate of degradation of water quality (Issue 11), and the rate of change in the composition of the atmosphere (Issues 5 and 6).

And finally, carrying capacity may ultimately have less to do with population density and more to do with a society's technology, resource demand, and waste.

---

[2] Simon, J., & H. Kahn. 1984. *The Resourceful Earth* (Malden, MA: Blackwell Publishers).
[3] Postel, S. 1996. *Dividing the Waters: Food Security, Ecosystem Health, and the New Politics of Scarcity* (Washington, DC: Worldwatch Institute).

# ASSESSING CARRYING CAPACITY

In mid-2012 the population of the United States was nearly 312 million.[4] Can our own natural resources successfully support 312 million people? How do the imported goods that maintain our level of economic activity and standard of living affect the global environment? How fast is our population growing? How much of our growth is due to legal immigration? Illegal immigration?

Obviously, the precise human carrying capacity of an area is very difficult to determine, especially for a country as large and diverse as the United States. Despite the difficulty, it is useful to estimate human carrying capacity so that policy makers can act to ensure that our environment is able to support human life and natural biodiversity into the future.

**Question 2-1:** Select one of the items in the following list and write a few sentences arguing that it is a carrying capacity issue, and then argue that it is not.

a. Die-offs of squirrels on your campus

b. The loss of 70,000 $km^2$ of cropland each year due to nutrient depletion

c. Increasing vehicular traffic and "road rage" incidents

d. Record high prices for heating oil in recent winters in the northeast United States and elsewhere

e. High real estate prices in California coastal communities

f. The loss of 50 percent of wetlands in twenty-two states since colonial times

**Question 2-2:** What indicators, environmental or otherwise, would support the assertion that the United States had exceeded its carrying capacity? One possible example: export of hazardous waste to developing countries. Support your selections.

It has been stated that everyone in the United States could fit comfortably inside the state of Texas. The mid-2012 United States population was 312 million. The area of Texas is 261,914 sq. mi. (67,835,000 ha).

**Question 2-3:** Calculate how many acres, and hectares, each person would occupy if all U.S. residents (in 2012) lived in Texas (1 $mi^2$ = 640 acres; 1 acre = 0.4 ha).

---

[4] U.S. Bureau of the Census, U.S. POPClock Projection www.census.gov/cgi-bin/popclock.

## ECOLOGICAL FOOTPRINT

Another way of examining the impact of humans is to estimate the individual impact each of us has, a measure known as our *ecological footprint.* Every individual has an ecological footprint that extends well beyond the geographical area in which that person lives. *The Ecological Footprint of Nations* gives estimates for how much of the Earth's area we appropriate for our "needs."[5] The average American in 2010, for example, used 8 ha (20 acres) to support his or her lifestyle. This includes farmland, forests, mines, dumps, schools, hospitals, roads, playgrounds, malls, etc. (see Figure 2-1).

**Question 2-4:** How many times larger would Texas need to be, assuming all Americans lived there and each American required 8 hectares?

**Question 2-5:** The surface area of the Earth is 15 billion hectares. Now assume all 7.05 billion people on Earth in 2012 lived like Americans. How much area would be needed, assuming all lived with the ecological footprint of Americans? Report your answer in hectares and acres. How many more planets with the surface area of Earth would be required?

**FIGURE 2-1** Aerial photo of a mall in Charleston, SC. To comply with local zoning regulations, shopping centers must have a minimum number of paved parking spaces, which contributes to the large ecological footprint of Americans. (Courtesy of Shutterstock.)

---

[5] Venetoulis, J., D. Chazan, & C. Gaudet, at http://www.redefiningprogress.org/newpubs/2004/footprintnations2004.pdf.

Of course, not all of the Earth's surface can serve as footprint area. Approximately 6.4 billion hectares are marginally productive or unproductive, since they are covered by ice (at least for the present) or lack water. Additionally, 36.3 billion hectares are covered with seawater.

**Question 2-6:** Summarize the major points of this Issue.

**Question 2-7:** Evaluate the carrying capacity of the United States from the standpoint of sustainability.

# FOR FURTHER THOUGHT

**Question 2-8:** Americans use a disproportionate portion of the Earth's resources. Discuss reasons why this is so.

**Question 2-9:** Discuss whether the use of resources by Americans is a fair allocation of the planet's resources.

**Question 2-10:** Is our use of resources sustainable? Why or why not? Cite specific examples and document your assertions with evidence.

**Question 2-11:** Because estimating an area's human carrying capacity is inexact and difficult, some would argue that the concept is useless. Do you agree or disagree? Explain and justify your answer.

**Question 2-12:** Conduct a search on the Web using the key words "ecological footprint." Once you have accessed a suitable site, calculate your footprint. How does your ecological footprint compare to others? What actions could you take to reduce your footprint?

**Question 2-13:** Go to the site of the Public Television series *Our Changing Planet,* (http://www.umac.org/ocp/CarryingCapacity/info.html). What are the various estimates of global human carrying capacity mentioned? Why do they vary from about 1 billion to 33 billion?

## COASTAL POPULATION GROWTH: BANGLADESH

### KEY QUESTIONS

- How does human population growth threaten coastal areas?
- How can we measure these threats?
- What can we do about them?
- Who is responsible for addressing the problem?
- To what extent, if any, is coastal population growth compatible with a sustainable society?

### INTRODUCTION: WHERE DO YOU LIVE?

Chances are you live within 100 kilometers (62 mi) of the coast. According to the World Resources Institute, at least 60 percent of the planet's population lives that close to the coastline.[1] Coastal areas have the fastest growing populations as well. Not surprisingly, over half the world's coastlines are at significant risk from development-related activities. Some of these activities are

- Destruction of tropical mangrove communities for fish and shrimp farms (see Issue 23).
- Expansion of many coastal cities that are in the direct path of hurricanes, monsoons, and other tropical storms.
- Pollution from Western-style industrial agriculture, including industrial meat production (Issue 18).
- Releases of untreated or partially treated human sewage: Between one-third and two-thirds of the human waste generated in developing countries is not even collected.

In addition, coastal cities from Miami to Mumbai are at increasing risk of flooding due to human-induced climate change.

Coastal cities have reached sizes unprecedented in human history. Here are but a few examples: São Paolo, Brazil, and Mumbai, India, each more than 19 million; Lagos, Nigeria, more than 20 million (and growing at over 5% per year); Dhaka, Bangladesh, over 16 million in 2011 (growth rate over 4% per year).

---

[1] Averaged for the period 1995–2000. World Resources Institute, www.wri.org.

# BANGLADESH

Bangladesh (Figure 3-1), a country of about 57,000 square miles (about the area of Illinois or Florida, or a quarter of the area of Saskatchewan), lies on the northern shore of the Indian Ocean. Dr. Selina Begum of the United Kingdom's Bradford University wrote:

> Bangladesh is a delta of the Ganges, Brahmaputra and Meghna Rivers. Tributaries and distributaries [branches] of the river system cover all of the country. The rivers rise in the Himalayas and drain a catchment area of about 1.5 million km$^2$ [577,000 mi$^2$], *only 7.5 percent of which lies in Bangladesh* [our emphasis].
>
> The country is prone to meteorological and geologic natural disasters, due to its geographic location, climate, variable topography, dynamic river system and exposure to the sea. The steady increase in population continuously increases the potential for natural disaster.[2]

Bangladesh thus makes an ideal case to illustrate the impacts of coastal growth. It was originally part of Pakistan after the Indian subcontinent wrested independence from Britain in 1947. In 1970, Bangladesh seceded from the rest of Pakistan. It had a UN-estimated 2011 population of 161 million. Bangladesh is vulnerable to catastrophic flooding from river discharges as well as from tropical storms. In addition, it is vulnerable to damage from sea level rise since more than half the country lies at an elevation of less than 8 meters (26 ft) above sea level (Figure 3-2). About 25 to 30 percent of the country is flooded each year, increasing to 60 to 80 percent during major floods. Moreover, more than 17 million people live on land that is less than 1 meter (3.3 ft) above sea level. River flooding in Bangladesh can result from some combination of the following: (1) excess rainfall and snowmelt in the watershed, especially in the foothills of the Himalayas (an area where many of the Earth's rainfall records were set); (2) simultaneous peak flooding in all three main rivers; and (3)

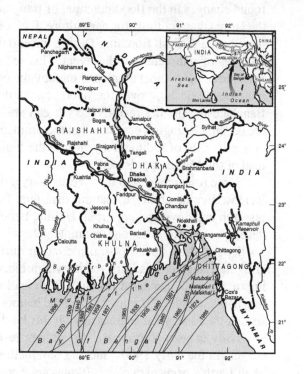

**FIGURE 3-1** Location of Bangladesh, showing tracks of major historical typhoons.

---

[2] Begum, S. 1996. Climate change and sea-level rise: Its implications in the coastal zone of Bangladesh. In *Global Change, Local Challenge: HDP* (Human Dimensions of Global Environmental Change Programme) Third Scientific Symposium, 20–22 September 1995; Geneva.

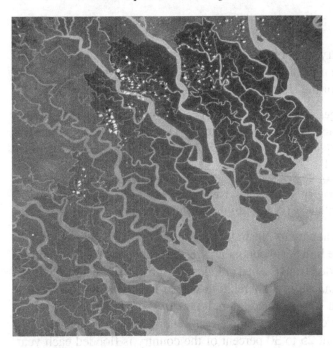

**FIGURE 3-2** High altitude photo of the Bangladesh coastal plain. The entire area is commonly under water during monsoon season. White areas are sediment plumes. (Courtesy of Stocktrek/Getty Images USA, Inc.)

high tides in the Bay of Bengal, which can dam runoff from rivers and cause the water to "pond up" and overtop its banks.

In addition, changes in land use in the far reaches of the watershed can result in profound changes in the flood potential of Bangladesh's rivers. For example, deforestation of the Himalayan foothills for agriculture, fuel, or human habitation in Nepal makes the land more prone to flash flooding. During torrential monsoon rains, mudslides can dump huge volumes of sediment into the rivers. This sediment can literally pile up in the river channels, reducing channel depth and thereby the channel's capacity to transport water. Result: more water sloshing over the rivers' banks during floods, and more disastrous floods in Bangladesh. Another result: The silt can smother offshore reefs when it finally reaches the ocean. Geologists estimate that the drainage system dumps at least 635 million tonnes ($1.4 \times 10^{12}$ lb) of sediment a year into the Indian Ocean, which is four times the sediment load of the Mississippi River.

Some researchers feel that Western-style development is contributing to Bangladesh's increased vulnerability to natural disasters. Farhad Mazhar, director of the Bangladeshi research and development group Ubinin, believes that the basic problem *is* Western-style development. He cites one example: During predevelopment days Bangladeshi villagers depended on boats for transportation during the monsoon season. But, "to be modern" the Bangladeshis built roads, which blocked floodwaters. These and other "modern improvements" increased flooding from cyclones with resultant damage and loss of life. Mazhar concludes, "If you compare the history of the old cyclones and flood waters, you'll see . . . the misery has increased with this kind of 'engineering' solution to water."[3]

Truly catastrophic coastal floods can result when high tides coincide with tropical storms. In the early 1990s, such a combination took over 125,000 lives. About one-tenth of all Earth's tropical cyclones (typhoons) occur in the Bay of Bengal (see Figure 3-1), and about 40 percent of the global deaths from storm surges occur in Bangladesh.

---

[3] Mazhar, F. What we want from Kyoto: OTN explores global warming. www.megastories.com/warming/bangla/kyoto.htm. 6 April 1998. [Unavailable, but see http://www.southsouthforum.org/eng/?page_id=922.].

*Storm surges*[4] result when cyclones with winds that can exceed 250 km/hr (155 mi/hr) move onshore, and push a massive wall of seawater onshore with them. Surges in excess of 5 meters (16.4 ft) above high tide are not uncommon, so you can imagine the impact such a storm can have on a country in which half the land is within 8 meters (26.2 ft) of sea level! Conservative computer models forecast an average rise of sea level of 66 centimeters (26 in) by 2100 due to global warming, but sea level could rise anywhere from 20 to 140 centimeters (8–55 in). There is, however, a complicating factor: The deltas on which Bangladesh is built are subsiding, meaning the area will be more susceptible to the flooding that will accompany sea-level rise.

*Subsidence* is a natural feature of deltas. The water in a large river may reach the sea in one or more branches, called distributaries. Over time, distributaries may meander more and more, thus lengthening the river's route to the sea. During floods, a distributary can overtop its banks and cut a new and shorter route to the sea. This can result in a rapid shift in the point of local sedimentation at the delta's mouth. An existing delta lobe can sometimes be quickly abandoned. Moreover, recall the sediment deposited over Bangladesh each year: since the muddy sediment of a delta is slowly compacting due to the weight of overlying sediment, without constant addition of new sediment, the delta subsides.

Here's Dr. Begum again:

A one-meter (3.3 ft) rise in sea level could turn a moderate storm into a catastrophic one. The tropical cyclones of the Bay of Bengal form only when sea surface temperatures reach 27°C or higher. Tropical cyclones in the Bay of Bengal generally form in the pre-monsoon (April–June), and post-monsoon (October–December) seasons. An increase in the surface air temperature (due to global warming) might lead to a more widespread occurrence of cyclones in the Bay of Bengal.[5]

Clearly, Bangladesh faces the potential for catastrophe from the combination of coastal population growth, coastal subsidence, land-use changes in the river watershed area, and global climate change (see Issues 5 and 6).

## POPULATION GROWTH IN BANGLADESH

To begin our analysis, we will *project* the population of Dhaka, Bangladesh, in the future using the doubling time formula, $t = 70/r$ (where t equals the doubling time and r equals the growth rate, expressed as a percent; see "Using Math in Environmental Issues" pages 6–8 for an explanation and example of how to use the doubling time equation.)

**Question 3-1:** The population of Dhaka was 16 million in 2011. Calculate the doubling time at the 2010 growth rate of 6 percent per year.

You can also project population growth using the compound growth formula. Although a city's or country's population size is a meaningful number, it becomes more meaningful when you know the *population density* at which that population lives. This is calculated by dividing the population by the area it inhabits.

---

[4] For details on storm surges go to the USGS home page at www.usgs.gov.
[5] Begum, op cit.

**Question 3-2:** Bangladesh had a 2011 population of 161 million on a land area of 144,000 km$^2$ (57,000 mi$^2$).[6] Calculate the nation's population density in persons per km$^2$ and per square meter.

**Question 3-3:** How do these numbers compare to the population density of your home state or province? State population densities are available at the U.S. Census Bureau's website.[7]

**Question 3-4:** Bangladesh's 2009 emissions of $CO_2$, the principal greenhouse gas, totaled 55.3 million metric tons, compared to the United States' 5.4 billion metric tons, and China's 7.7 billion, which was by far the world's major source. The per capita emissions were even more disparate: 0.36 metric tons per Bangladeshi compared to 17.7 metric tons for each resident of the United States, and 5.8 tonnes for each Chinese.[8] What was the ratio between U.S. per capita emissions in 2009 and those of Bangladesh?

**Question 3-5:** Summarize the major points of this Issue.

---

[6] CIA World Factbook, www.cia.gov.

[7] You can go to www.census.gov and search for "state densities."

[8] See for example http://www.guardian.co.uk/news/datablog/2011/jan/31/world-carbon-dioxide-emissions-country-data-co2#data.

# FOR FURTHER THOUGHT

**Question 3-6:** What responsibility, if any, do you believe the United States has to mitigate (lessen) the potential for environmental disaster in Bangladesh?

**Question 3-7:** Assume that the United States and China are at least partially responsible for the potential for environmental disaster in Bangladesh. List actions, in order of importance, that the United States could take to prevent, postpone, lessen, or clean up environmental damage and resultant destruction of property and loss of life.

**Question 3-8:** What are some of the implications of human population growth for the coastal environment in general? Consider air and water quality, impervious surface, land use, and species loss. Explain why these are or are not causes for concern.

# Issue 4

# POPULATION GROWTH AND MIGRATION

## KEY QUESTIONS

- What are fertility rates and how are they determined?
- How much does migration contribute to population growth in the United States?
- Can the environmental impacts of immigration be quantified?
- How is migration related to the concept of sustainability?

## INTRODUCTION: POPULATION GROWTH

Global population growth rates in 2012 ranged from an estimated –1.0 percent in Moldova to +4.9 percent in Qatar.[1] What factors contribute to these growth rates? Are growth rates rising or falling? Is a growth rate of 0.8 percent high or low? We will address these and similar questions in this issue.

The United States' 2012 growth rate of 0.9 percent would result in the population doubling to more than 600 million in about 78 years (see "Using Math in Environmental Issues," pages 6–8 for a discussion of doubling time). Among industrialized countries, this is a high rate. For example, the 2012 population growth rates were –0.2 percent in Germany, -0.48 percent in Russia, 0.50 percent in France, and 0.55 percent in the United Kingdom.[2] Overall, world population growth rates are declining, although global population, at 1.096% per year in 2012, continues to grow.[3]

For a population to increase, there must be more births than deaths and/or greater immigration than emigration. Let us first examine births, which demographers refer to as *natality* or *fertility.* The crude birth rate (CBR), is the number of births per 1,000 people and is found by dividing the total births for the year by the mid-year population, then multiplying by 1000. Globally, 2012 CBRs range from lows of 7.5 in Singapore, 8.3 in Germany, and 9.0 in Italy, to highs of 39 in Afghanistan, 40 in the Democratic Republic of the Congo, and 50 in Niger. The 2012 crude birth rate for the United States was 13.7 live births per 1,000 people,[4] which is somewhat higher than other industrialized countries but low compared with the developing world. Population growth can be projected using crude birth and death rates.

---

[1] CIA World Factbook www.cia.gov.

[2] Ibid.

[3] United Nations Population Division (http://www.un.org/esa/population/unpop.htm).

[4] CIA World Factbook, op cit.

**Question 4-1:** Project growth in the U.S. population at the 2012 crude birth rate of 13.7 and crude death rate of 8.4 by completing the following table, starting at the 2012 population (309,000,000). For this calculation we will ignore the data for migration.

| Year | Initial Population | Net Addition to Population (to nearest hundred) | Total Population at End of Year |
|------|--------------------|------------------------------------------------|---------------------------------|
| 2012 | 309,000,000 | 1,637,700 | 310,637,700 |
| 2013 | 310,637,700 | | |
| 2014 | | | |
| 2015 | | | |
| 2016 | | | |
| 2017 | | | |

To make this calculation, first subtract the crude death rate from the crude birth rate. (This quantity is known as the *natural increase*). Then multiply this number by the beginning population divided by 1000. (NOTE: Some calculators will not allow you to enter numbers this large without using scientific notation.) This quantity allows you to calculate the natural rate of population growth, that is, growth that doesn't count migration.

**Question 4-2:** Use the formula $r = (1/t)\ln(N/N_0)$ (see "Using Math in Environmental Issues," pages 6–8) to calculate the annual population growth rate of the United States using data from the above table. (Recall that this rate ignores the data for migration.)

**Question 4-3:** Assume a stable population to be our goal, then suggest policies that might reduce the birth rate in the U.S. Which policies are the most feasible? Why? Which are the least feasible? Why? Is our assumption reasonable? Why or why not?

Although both the crude birth and death rates are simple to calculate and easy to use, a better measure of fertility is *total fertility rate* (TFR), the average number of children that a woman will bear over her reproductive life. Again ignoring migration, for the U.S. population to stabilize the TFR should equal about 2.1.[5]

---

[5] A TFR of 2.1 allows for a low rate of infant and child mortality. Thus, in countries with high infant and child mortality rates, the replacement TFR is substantially higher.

## MIGRATION AND POPULATION GROWTH

In the preceding questions we calculated the annual population growth rate of the United States given the 2012 crude birth and death rates and ignoring migration. Now let's calculate the rate of growth of the U.S. population using the data for migration. Emigration (outmigration) is relatively small (~200,000 annually); net immigration was estimated at 3.62 per 1,000 persons in 2012.[6]

**Question 4-4:** Calculate the number of legal immigrants to the U.S in 2012 (net migration = 3.62 per thousand population).

The number of immigrants illegally entering the country is difficult to estimate (Figure 4-1). Conservatively, many experts estimate illegal immigration during 2009–2012 to be between 300,000 and 500,000 per year, a significant decline from the rates of 2000–2007. The TFR of new immigrants (72) is significantly higher than that of residents who are not recent immigrants (62).

**Question 4-5:** Use your calculator or computer spreadsheet program to fill in the table below, assuming the previous crude birth and death rates and net immigration of 1.4 million (legal + illegal) per year.

| Year | Initial Population | Net Addition — Births Minus Deaths | Net Addition — Immigration | Population at End of Year |
|------|-------------------|-----------------|-------------|---------------------------|
| 2012 | 309,000,000 | 1,637,700 | 1,400,000 | 312,037,700 |
| 2013 | 312,037,700 | | 1,400,000 | |
| 2014 | | | 1,400,000 | |
| 2015 | | | 1,400,000 | |
| 2016 | | | 1,400,000 | |
| 2017 | | | 1,400,000 | |

**FIGURE 4-1** Haitians crowd a boat at Port-au-Prince in an attempt to reach the United States. (Courtesy of David Turnley/Corbis Images.)

---

[6]  U.S. Dept. of Homeland Security Yearbook of Immigration Statistics: 2003 (http://www.dhs.gov/xlibrary/assets/statistics/yearbook/2003/2003IMM.pdf).

**Question 4-6:** Calculate the annual population growth rate from the preceding table using the formula $r = (l/t)\ln(N/N_0)$.

**Question 4-7:** For 2012, what percentage of total population growth was due to immigration? Explain whether you think that immigration makes a significant contribution to U.S. population growth.

**Question 4-8:** List the positive contributions that you believe immigrants make to the United States. Identify any assumptions you used (i.e., diversity is desirable, etc.).

A 1997 report by the National Research Council at the request of the U.S. Commission on Immigration Reform attempted to quantify the impact of immigration on U.S. society. Among its conclusions were the following:[7]

- If immigration continued at present (1997) levels, it would account for two-thirds of U.S. population growth to 2050.
- Immigrants are more poorly educated than residents, thus imposing significant costs on localities.
- Immigration lowers wages among less-skilled residents and is responsible for nearly 50 percent of wage depression among low-skilled U.S. workers since 1980.
- Immigration is a contributing factor in the widening gap between the rich and the poor in the United States.
- Immigration imposes considerable cost on U.S. society with little economic benefit. Immigrants may add $1 to $10 billion annually to U.S. gross domestic product (GDP; the total annual value of goods and services), but they cost $15 to $20 billion per year in government services.

**Question 4-9:** Given that immigrants need greater education services than others, should the federal government (which controls immigration) assist states and localities with these additional costs? Why or why not?

---

[7] National Research Council. 1997. *The new Americans: Economic, demographic, and fiscal effects of immigration* (Washington, DC).

Thus, according to the NRC report, at a time of flat or reduced outlays for social services, over 1 million immigrants per year apparently impose significant costs on U.S. society, with much of that cost falling on localities, and those least able to pay.

**Question 4-10:**  For an alternate view on the impact of illegal immigration, go to http://www.npr. org/templates/story/story.php?storyId=5312900. Using both the NRC and NPR sources, discuss ways in which the United States might be affected by reducing immigration. List some advantages and disadvantages (e.g., economic, social, political, and cultural gains/losses) and justify your choices.

**Question 4-11:**  How might migration be part of a sustainable society?

**Question 4-12:**  Summarize the major points from this Issue.

# FOR FURTHER THOUGHT

**Question 4-13:**  From an environmental perspective, immigrants to the United States quickly begin to consume energy and natural resources at the same rate as residents. Some assert that immigration to the United States should be limited in order to conserve energy and natural resources. Is it reasonable to support immigration reductions for this reason without seeking to reduce Americans' own disproportionate consumption of global resources?

**Question 4-14:**  Research the impact of immigration on Germany, Austria, Italy, France, Spain, Israel, and Greece, and describe how these countries are responding.

# Issue 5

## GREENHOUSE GASES AND CLIMATE CHANGE: PART ONE

### KEY QUESTIONS

- What is the composition of the Earth's atmosphere?
- How has the atmosphere changed over planetary history?
- What processes influence and alter the atmosphere's composition?
- What are greenhouse gases, and how do they affect global temperatures?
- How is global sustainability affected by a changing climate on the scale forecasted by atmospheric scientists?

### FACTS ABOUT THE ATMOSPHERE

All life on Earth depends on our atmosphere. Despite this, humans are altering the atmosphere's composition with growing understanding, at least on the part of scientists, of medium- to long-term adverse impacts. Scientific evidence has confirmed that emissions from the burning of fossil fuels, from industrial sources such as cement manufacture, and from deforestation have changed and continue to alter the makeup of our atmosphere. In addition, trace gases such as methane and chlorofluorocarbons (CFCs and HCFCs) are having an impact on the atmosphere wholly out of proportion to their concentration. To analyze the effects of human activity on the atmosphere, there are some facts you need to know.

Earth's atmosphere is a thin shell. Half the mass of the atmosphere is within 5 km (3.1 mi) of the surface, and 90 percent is within 20 km (12.4 mi). Twenty km is not very far—about a 10-minute drive on most interstate highways. By comparison, the Earth's radius is 6,378 km (3963 mi).

**Question 5-1:** What percent of the Earth's radius is 20 kilometers?

The major gases that make up the Earth's atmosphere are shown in Table 5-1. The proportions of the gases (ignoring water vapor, which is variable) in the atmosphere remain virtually constant to a height of about 25 km.

TABLE 5-1 ■ Major Gas Composition of the Atmosphere

| Gas | Volume (%) | Molecular Weight (approximate) | Weight (%) |
|---|---|---|---|
| Nitrogen | 78.000 | 28 | 75.00 |
| Oxygen | 21.000 | 32 | 23.00 |
| Argon | 00.930 | 40 | 1.30 |
| Carbon dioxide | 00.035 | 44 | 0.05 |
| Average atmosphere | 100.000 | 29 | 100.00 |

*Percentages do not add up to 100 due to rounding, approximation, and ignoring water vapor and all minor and trace gases.
Weight (%) is found by multiplying the volume (%) by the molecular weight of that gas and dividing by the average molecular weight of the atmosphere.

## WEIGHING THE ATMOSPHERE

At sea level, the weight of a 1 cm$^2$ column of air is approximately 1 kilogram. This value, 1 kg/cm$^2$, is also known as the atmospheric *pressure*. For this analysis, we will assume a featureless Earth without continents. This will simplify, but not significantly alter, your calculations.

**Question 5-2:** How much does the atmosphere weigh over each square meter of surface? Express this in tonnes (units of 1000 kg). Recall the conversion factors: $10^4$ cm$^4$ = 1 m$^4$; 1 tonne = 1000 kg = $10^6$ g.

**Question 5-3:** How much does the atmosphere weigh over each square kilometer (km$^2$)? Recall that $10^6$ m$^2$ = 1 km$^2$.

**Question 5-4:** The area of the Earth's surface is approximately $516 \times 10^6$ km$^2$. What is the weight of the entire atmosphere in tonnes? Express your answer using scientific notation, in units of $10^{12}$ tonnes.

It is worth pointing out that this is by no means an incomprehensibly large number. The mass of the Earth is $6 \times 10^{27}$ g. How does the atmosphere's mass compare with that of the Earth?

## EARLY PLANETARY ATMOSPHERES

Our current atmosphere is believed to be the Earth's third over its cosmic history. The earliest was comprised of light gases, primarily hydrogen ($H_2$) and helium ($He_2$). This first atmosphere was blown away as the Sun reached critical mass and began internal fusion, and by heat generated by the early molten earth, between 4 and 5 billion years ago. The second atmosphere was formed around 4 billion years ago when gases escaped from the Earth as its crust solidified. Bombardment by comets may have also contributed gases to the Earth's second atmosphere. This second atmosphere lacked free oxygen, a gas necessary to support most of the Earth's present life forms.

## THE DEVELOPMENT OF THE CURRENT ATMOSPHERE

The Earth's current atmosphere was produced partly by the metabolism of living organisms. Cyanobacteria, primitive photosynthesizing organisms, had appeared by 3.8 billion years ago and began to produce oxygen as a byproduct. Over the next 2 billion years, atmospheric oxygen ($O_2$) concentrations rose and $CO_2$ concentrations fell as a direct result of photosynthesis, as well as the sequestration (storage or "fixing") of carbon in *carbonate* rock (limestone and dolostone: $CaCO_3$ and $CaMg[CO_3]_2$). This increase in atmospheric oxygen set the stage for two major evolutionary events on the planet: the evolution of aerobic (oxygen-using) life forms and the formation of the *ozone layer.*

Increasing oxygen in the hydrosphere and atmosphere was toxic to sensitive *anaerobic* microorganisms and prompted the evolution of new microorganisms capable of more efficient respiration, using oxygen to liberate energy from organic compounds. Today, large anaerobic microbial communities are restricted to Mid-Ocean Ridge hydrothermal systems and other marginal environments. The development of aerobic metabolism also permitted the evolution of multicellular organisms, which required more energy to support their increased biomass. All existing multicellular life forms employ aerobic metabolism.

Increasing atmospheric oxygen reacted with high-energy solar radiation and eventually triggered the formation of a layer of ozone ($O_3$) in the upper atmosphere. Although a pollutant at ground level, the ozone layer filters out harmful high-energy ultraviolet radiation, which can cause skin cancer in humans.

Indeed, before the formation of the ozone layer, the Earth's surface was exposed to extremely high intensities of ultraviolet radiation. The surface ocean waters filtered out some of this radiation and thus provided some protection to organisms, but it is likely that ocean primary production (that is, production of high-energy compounds from photosynthesis) was still limited by the high ultraviolet radiation intensities. The terrestrial surface fared worse and was possibly even sterilized by this radiation: The first widely accepted evidence for land plant communities, for example, appears in rocks that are only about 400 million years old.

## FUNCTIONS OF THE ATMOSPHERE

In addition to providing the oxygen needed by most of Earth's life forms, the atmosphere provides thermal insulation, preventing extreme changes in temperature over the daily light-dark cycle. Unequal heating of the Earth's atmosphere and terrestrial surface create long-term *climate* and short-term *weather* patterns. The winds that result from these heating differences and resultant pressure differences help drive ocean currents. The atmosphere also transfers vast quantities of heat from equatorial to polar latitudes.

# THE ATMOSPHERE'S CURRENT COMPOSITION

As Table 5-1 shows, the present-day atmosphere is composed primarily of nitrogen ($N_2$) gas (78.08% by volume), oxygen ($O_2$; 20.94%), and argon (Ar; 0.93%). It also contains water vapor ($H_2O$), carbon dioxide ($CO_2$), neon ($Ne_2$), helium ($He_2$), methane ($CH_4$), oxides of sulfur (SOx) and nitrogen (NOx), and ozone ($O_3$). The major controls on the composition of the atmosphere, and the cycling of these compounds in and out of the atmosphere, are interactions with the Earth's biosphere (living matter), hydrosphere, and lithosphere (rock and geological processes such as volcanism).

Presently, $O_2$ levels seem to be stable, but $CO_2$ levels are not and display seasonal and longer-term trends. Seasonal changes in $CO_2$ are related to seasonally fluctuating primary production (i.e., plant growth) due to changing light durations. Longer-term (decade-to-century) increases in $CO_2$ are due to a variety of anthropogenic (human-caused) inputs, including changes in land use that reduce the ability of terrestrial biota to absorb $CO_2$. The present concentration of $CO_2$ is higher than it has been for at least 400,000 years, according to the Intergovernmental Panel on Climate Change (IPCC).

Because the amount of $CO_2$ in the atmosphere is very small, the concentration is easily changed by the addition of $CO_2$ from various sources.

# CHANGES CAUSED BY HUMANS

After decades of research, scientists have concluded that humans have altered the composition of the Earth's atmosphere. Before the Industrial Revolution (ca. 1750), clear-cutting of forests in Europe, China, and the Middle East, and later in North America, set the stage for modifying the atmosphere's composition. Cutting and burning forests liberates $CO_2$ in two ways. Carbon from the vegetation is converted to $CO_2$ and soils, devoid of their tree cover, emit $CO_2$ at greater rates than before.

Since the beginning of the Industrial Revolution, atmospheric concentrations of greenhouse gases have increased. Carbon dioxide has risen by one third; methane has more than doubled; and nitrous oxide concentrations have risen by about 15 percent (Table 5-2).

## The Greenhouse Effect

Here's how the *greenhouse effect* works. Greenhouse gases allow short wavelength radiation (mainly visible light and ultraviolet) from the Sun to pass through the atmosphere, but they absorb the longer wavelength radiation (mainly infrared) that is emitted by the Earth's surface, thereby heating the atmosphere. Over the past two centuries, (1) the global human

**TABLE 5-2** ■ Changes in the Global Concentration of Greenhouse Gases Since the Preindustrial Period[1]

|  | $CO_2$ | $CH_4$ | $N_2O$ |
|---|---|---|---|
| Preindustrial concentration | 280 ppmv | 700 ppbv | 275 ppbv |
| Concentration in 2005 | 380 ppmv | 1720 ppbv | 312 ppbv |
| Concentration in 2011 | 391 ppmv | 1810 ppbv | 332 ppbv |
| Atmospheric lifetime (years) | 50–200 | 12 | 120 |

ppmv = per million by volume; ppbv = per billion by volume

[1]Intergovernmental Panel on Climate Change (IPCC). 1995: The science of climate change: Contribution of Working Group I to the Second Assessment Report to IPCC on Climate Change. J.T. Houghton, L.G. Meira Filho, B.A. Callander, N. Harris, A. Kattenberg & K. Maskell, eds. (Cambridge University Press: New York). Courtesy of J. T. Houghton, L. G. Meira Filho, B. A. Callander, and N. Harris/IPCC Secretariat.

Reference: http://cdiac.ornl.gov/pns/current_ghg.html, Oak Ridge National Lab.

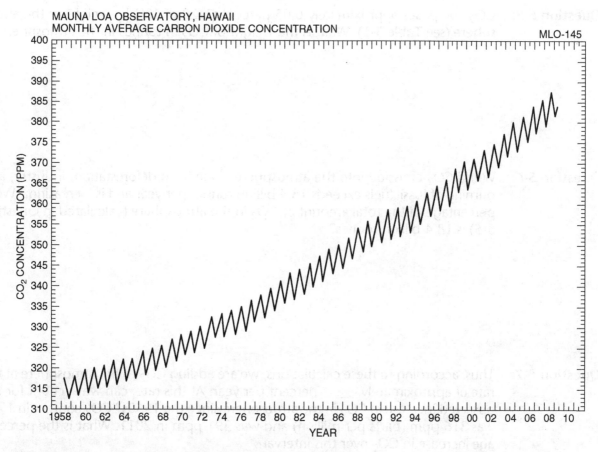

MAUNA LOA OBSERVATORY, HAWAII
MONTHLY AVERAGE CARBON DIOXIDE CONCENTRATION
MLO-145

**FIGURE 5-1**    Atmospheric $CO_2$ concentrations measured at the Mauna Loa Observatory from 1958 to 2010. (Source: C.D. Keeling and T.P. Whorf. Atmospheric carbon dioxide record from Mauna Loa, http://cdiac.esd.ornl.gov/trends/co2/sio-mlo.htm and http://cdiac.ornl.gov/pns/current_ghg.html)

population has grown tenfold; (2) the demand for energy to support industrial development, heat homes, cook food, watch television, and surf the Internet has vastly grown; and (3) the increased use of automobiles has resulted in the burning of great stores of fossil fuel. Figure 5-1 shows what is known as *the Keeling Curve*, the change in atmospheric $CO_2$ concentration over time.

Fossil fuels, including coal, oil, and natural gas, were formed by the burial and slow anaerobic decomposition of ancient plant and phytoplankton deposits. While these deposits took tens of millions of years to form, we are burning them at a vastly more rapid rate. A key by-product of fossil fuel consumption is $CO_2$. Since the Industrial Revolution, we have added $CO_2$ to the atmosphere far more rapidly than it can be absorbed by its variety of sinks, or "storehouses." This has led to a steady increase in $CO_2$ concentration that will result in at least a doubling of pre-1860 atmospheric $CO_2$ content by the year 2150, if present trends continue.

Methane is another greenhouse gas whose concentration has increased because of human activities. Methane is emitted by burping cows, flooded farmlands (i.e., rice paddies), coal seams, leaking gas pipelines, and municipal landfills.

## CARBON DIOXIDE IN THE ATMOSPHERE

Previously, you calculated the weight of the entire atmosphere. Now let's figure out how much of this weight is $CO_2$.

**Question 5-5:** $CO_2$ comprises approximately 0.05 percent (1/20 of 1%) by weight of the atmosphere (see Table 5-1). What is the weight of $CO_2$ in tonnes in the atmosphere?

**Question 5-6:** World $CO_2$ emission into the atmosphere alone from deforestation, industry, and burning of fossil fuels exceeds 18.4 billion tonnes per year and is increasing.[1] What percentage of the total amount of $CO_2$ in the atmosphere (calculated in Question 5-5) is 18.4 billion tonnes?

**Question 5-7:** Thus, according to these calculations, we are adding $CO_2$ to the atmosphere at the rate of approximately _____ percent per year. At this rate, can we account for the observed increase in $CO_2$ in the atmosphere since the 1950s? $CO_2$ content in 1959 was 316 ppm (parts per million) and was 391 ppm in 2011. What is the percentage increase in $CO_2$ over this interval?

**Question 5-8:** Is this percentage increase within the order of magnitude suggested by your calculations? (Order of magnitude means roughly within the same power of 10; e.g., the ratios 3.5 cm/yr and 6.5 cm/yr are within the same order of magnitude, whereas 3.5 cm/yr and 26.8 cm/yr are not.)

**Question 5-9:** Recall the introduction to this Issue: Do you think all of the $CO_2$ produced by human activities stays in the atmosphere? Identify a major "sink," or storehouse for this anthropogenic (human-generated) $CO_2$.

---

[1] http://co2now.org/ and www.iea.org.

In the next Issue, we will evaluate the nature of the climate change risk we face and what, if anything, we can do to mitigate it. For now, think about the importance of the rate at which climate changes.

**Question 5-10:**   Summarize the major points of this Issue.

**Question 5-11:**   Discuss the issue of greenhouse gases and climate change from the standpoint of sustainability.

## FOR FURTHER THOUGHT

**Question 5-12:**   Research the carbon cycle. Discuss the importance of limestone and fossil fuels as carbon storehouses (sinks).

**Question 5-13:**   Many countries with sizable populations living in coastal areas, as well as some small countries with their entire land area near or at sea level, are, or should be, acutely concerned about the impact of sea-level rise associated with human alteration of the atmosphere. The Maldives are one such country, and Bangladesh was discussed in Issue 3. But the countries with most to lose are probably China, Japan, and perhaps Vietnam, with hundreds of millions of people and trillions of dollars of real estate at risk by 2080. How are China, Japan, and Vietnam dealing with the risks of climate change?

**Question 5-14:**   Research the impact of "Superstorm" Sandy (October 2012) upon the Northeastern United States. Why was this storm so dangerous? Do climate scientists think that the storm could have been caused by climate change? Explain your answer.

# GREENHOUSE GASES AND CLIMATE CHANGE: PART TWO

## KEY QUESTIONS

- How fast is the Earth's climate changing?
- What impacts of global climate change will result from greenhouse gas increases?
- How does the Earth's atmosphere interact with the ocean?
- How can climate change affect ocean circulation?
- Can human-induced global climate change be reversed?

During the summer of 2012, thousands of temperature records were broken across the United States from the Rocky Mountains to the Eastern Seaboard. An expected bumper corn crop was decimated by extreme heat and drought. The West Coast temperatures were normal. Are these harbingers of a new normal?

**Question 6-1:** Access http://www.globalchange.gov/images/cir/hi-res/11-southeast-pg-112_top.png, an official government forecast of the increase in very hot days in the American Southeast due to human-induced climate change. How many more days above 90°F is eastern Virginia likely to experience?

## INTRODUCTION: THE 3RD IPCC ASSESSMENT

At a meeting in Shanghai in January 2001, the Intergovernmental Panel on Climate Change (IPCC) issued its most comprehensive report to date on the environmental implications of climate change. Over 150 delegates from nearly 100 governments met to consider the Third Assessment Report of the IPCC "Climate Change 2001: The Scientific Basis."[1] The full report, which runs to over 1,000 pages, is the work of 123 lead authors, who used the contributions of more than 500 scientists. The report went through extensive review by experts and governments. In line with IPCC Principles and Procedures, after line-by-line consideration, the governments unanimously approved the Summary for Policymakers of the report and accepted the full report.

---

[1] Available at www.grida.no/climate/ipcc_tar/.

Here are the major conclusions of this report:[2]

1. Confidence in models of future climate has increased. Climate data for the past 1,000 years, as well as model estimates of natural climate variations, suggest that there is an anthropogenic "signal" in the climate record of the last 35–50 years, meaning this change has certainly been affected by human activities. Analyses of tree rings, corals, ice cores, and historical records for the Northern Hemisphere indicate that the increase in temperature in the twentieth century is likely to have been the largest of any century during the past 1,000 years. It is likely that the 1990s were the warmest decade and 1998 was the warmest year during the past millennium.

2. In the mid- and high-latitudes of the Northern Hemisphere snow cover has likely decreased by about 10 percent since the late 1960s. The annual duration of ice cover shortened by about two weeks over the twentieth century. It is likely that there has been about a 40 percent decline in Arctic sea-ice thickness during late summer to early autumn in recent decades.

3. Since 1750, the atmospheric concentration of carbon dioxide increased from 280 parts per million (ppm) to about 367 ppm in 1998.[3] *The present $CO_2$ concentration has not been exceeded during the past 420,000 years and likely not during the past 20 million years* [our emphasis].

4. The global averaged surface temperature is projected to increase by 1.4–5.8°C from 1990 to 2100. This is higher than the 1995 Assessment Report's[4] projection of 1–3.5°C, largely because future sulfur dioxide emissions (which help to cool the Earth) are now expected to be lower—the result of lower air pollution. (However, during the 2000s particulate emissions from massive Chinese coal plants may have started to cool the Earth by reflecting and diffusing incoming solar radiation—a phenomenon not considered in early climate models.) This future warming is on top of a 0.6°C increase since 1861.

5. Global average water vapor concentration (humidity) and precipitation are projected to increase. More intense precipitation events are likely over many Northern Hemisphere mid- to high-latitude land areas. There is renewed debate on the subject after the 2005–2006 hurricane seasons.

6. Sea level was projected to rise by 0.09 to 0.88 meters from 1990 to 2100.

# UPDATE: THE 2007 (4TH) IPCC ASSESSMENT AND BEYOND

The 2007 IPCC Assessment[5] largely confirmed the findings of the 3rd Assessment, with these new findings:

1. The warming of the climate system is now "unequivocal."
2. Of the twelve years between 1995–2006, eleven were among the warmest ever measured (since 1850).
3. Sea level rise since 1993 was estimated at 2.4–3.8 mm/y, with an average rate of 3.1mm/y.

---

[2] Adopted with changes from "New Evidence Confirms Rapid Global Warming, say Scientists." UNEP News Release 01/5. Available at www.unep.org/Documents/Default.asp?DocumentID=296.

[3] Keeling, C.D. & T.P. Whorf. Atmospheric carbon dioxide record from Mauna Loa, http://cdiac.esd.ornl.gov/ trends/co2/sio-mlo.htm. By 2012 $CO_2$ had reached 391 ppm.

[4] Intergovernmental Panel on Climate Change (IPCC). 1995. Climate Change 1995: The science of climate change: Contribution of Working Group I to the Second Assessment Report of the Intergovernmental Panel on Climate Change. J.T. Houghton, L.G. Meira Filho, B.A. Callander, N. Harris, A. Kattenberg & K. Maskell, eds. (Cambridge University Press: New York).

[5] http://www.ipcc.ch/.

4. Average Northern Hemisphere temperatures during the last half of the 20th century were likely higher than at any time in the past 1300 years. There is "medium confidence" that "other effects" of climate change are emerging, including heat wave mortality in parts of Europe, and the expansion of the range of formerly tropical disease vectors.

5. The West Antarctic Ice Sheet contains enough ice to raise sea level by 5–6 meters. There is a "small chance" that the collapse of this sheet could occur during the next few centuries.

There is a scientific consensus that the principal drivers of current, accelerated climate change are burning of fossil fuels and deforestation. For the latest information, start at the following web sites: U.S. EPA Climate Change (http://www.epa.gov/climatechange/), NASA Global Climate Change (http://climate.nasa.gov/), and the Intergovernmental Panel on Climate Change (www.ipcc.ch/).

**Question 6-2:** Visit any of the above web sites and summarize current knowledge on the impacts and science of climate change.

## IMPACTS OF GLOBAL CLIMATE CHANGE

Examine Figure 6-1 for an overview.

1. *Human health* will be directly affected by increases in the *heat index*, which especially affects the elderly and those with heart and respiratory illnesses. (See above.) Higher temperatures will also increase ozone pollution in the lower atmosphere, a further threat to people with respiratory illnesses.

2. Warming may increase the incidence of some infectious diseases, particularly those that usually appear only in warm areas. Diseases that are spread by "vectors"—mosquitoes and other insects, etc.—such as malaria, dengue fever, yellow fever, and encephalitis, could become more prevalent if warmer temperatures and wetter climates enable those insects to become established farther poleward. Already, some municipalities are planning increased spraying of insecticides to combat tropical vector-borne (mainly by mosquitoes) diseases.

3. Alterations in precipitation patterns are likely, and in fact may already be occurring according to the 4th Assessment. Changes in climate are expected to enhance both evaporation and precipitation in most areas of the United States. The net balance of evaporation and precipitation influences the availability and quality of water resources. In areas expected to become more arid, like California, lower river flows and lower lake levels could impair navigation, reduce hydroelectric power generation, decrease water quality, and reduce the supplies of water available for agricultural, recreational, residential, and industrial uses. In other areas, increased precipitation is expected to be more concentrated in large storms as temperatures rise. This could increase the incidence and severity of flooding.

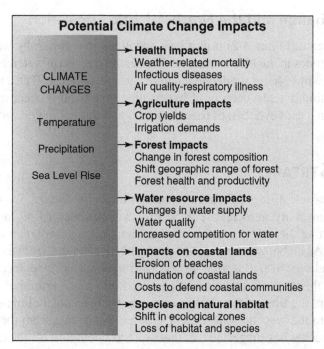

**FIGURE 6-1** Impacts of climate change.

4. Climate change could have severe impacts on agriculture (e.g., increasing evaporation from soils, thus drying them out). These impacts could be partially offset, at least in the United States, by longer growing seasons and enhanced crop production from higher atmospheric $CO_2$.

5. Climatic change will cause a shift in biotic community composition as plant and animal species try to migrate to maintain their preferred habitats. For example, the projected *average* 2+°C (3.6°F) warming likely to occur this next century could shift the ideal range for many North American forest species by about 300 kilometers (200 mi) to the north. Many plant species lack seed dispersion mechanisms that can adjust this rapidly, and these species may become regionally extinct. Coastal wetland plant communities may lose habitat because they may not be able to keep up with the predicted rate of sea-level rise, or the migration paths of these communities may be blocked by human development. Coral reef ecosystems, already under stress from human impact, may be reduced to a fraction of their historical range.

Wildlife species may be severely impacted by climate change[6]. For example, some duck species are dependent upon "prairie potholes" found in the Northern Great Plains. A drier climate would decrease the number and size of ponds in this region, with a commensurate reduction in duck populations. Fish that inhabit inland aquatic environments will be more vulnerable than coastal or marine species. Lake-locked fishes have little recourse in seeking cooler waters. Fish that inhabit north-south rivers may be able to migrate to seek cooler water, but fish in east-west oriented rivers and lakes will not be able to escape warming impacts. The diversity of fishes in U.S. rivers and streams is likely to decline, according to one study[7] in some local systems as much as 75%.

---

[6] Bellard, C. et al. 2012. Impacts of climate change on the future of biodiversity. *Ecol. Lett.* 15: 365–377.

[7] Xenopoulos, M.A. et al. 2005. Scenarios of freshwater fish extinctions from climate change and water withdrawal. *Global Change Biol.,* 11, 1557–1564.

### Climate Change and the North Atlantic Circulation

The Gulf Stream (Figure 6-2) brings vast quantities of warm, salty water from the tropics to higher latitudes in the North Atlantic. The heat in this warm water, coupled with prevailing westerly winds, helps moderate much of Europe's climate. The entire system seems to depend on rapid, intense cooling of hypersaline (extra-salty) water off Greenland, which sets in motion a conveyor belt of oceanic circulation throughout the world's oceans (Figure 6-3).

## A DISAPPEARING GULF STREAM?

In a number of scientific papers in such journals including *Nature, Science,* and *GSA Today,* Columbia University geoscientist W.S. Broecker has warned of "an oceanic flip-flop." Global warming, he argues, could interrupt the Gulf Stream's *thermohaline* circulation. Here's how: As global warming melts freshwater glaciers in Greenland, salt concentrations in surface waters will fall, leading to less mixing of surface and deep waters. This would stop or slow the thermohaline circulation (also known as *meridional overturning circulation*), interrupt the flow of the Gulf Stream, and bring a cooler climate to northern Europe as winds from the west no longer carry the heat emanating from the Gulf Stream. More recently, Broecker has suggested that these effects could turn off the deep-ocean conveyor belt completely, possibly triggering intense cooling in Europe. Evidence for such flip-flops has been found in geological records obtained from ice cores and deep-sea sediments. Of particular concern is the fact that these events have occurred over time periods as short as *four years.* Broecker calls the oceans "the Achilles' heel" of the climate system.

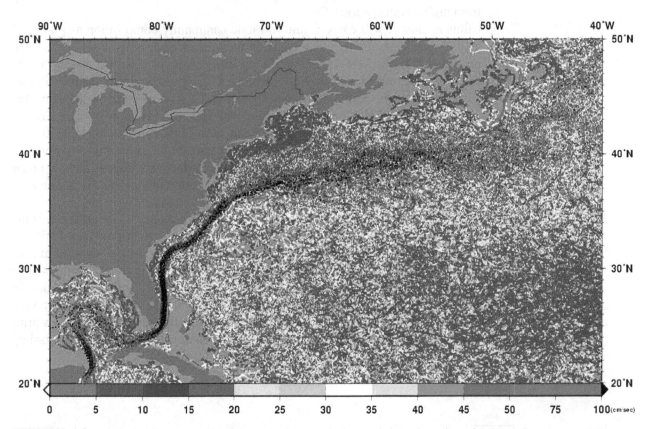

**FIGURE 6-2** The Gulf Stream. (Professor Arthur J. Mariano Phd, University of Miami, http://oceancurrents. rsmas.miami.edu. Courtesy of Rosenstiel School of Marine and Atmospheric Science.)

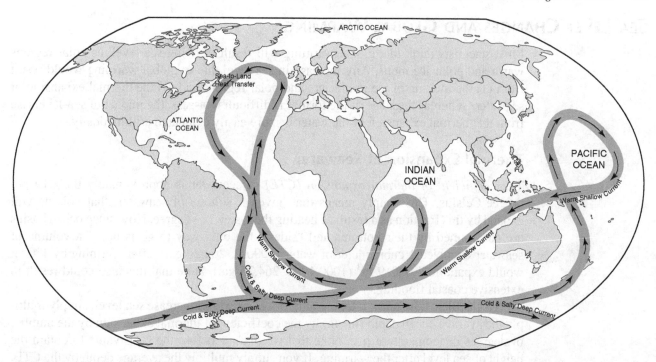

**FIGURE 6-3** Conveyor-belt circulation. This circulation is initiated when cool North Atlantic water sinks. It then flows southward and moves into the Indian and Pacific Oceans.

**Question 6-3:** In a 1999 paper in *GSA Today*, Broecker says, ". . . an extreme scenario is an unlikely one, for models suggest that in order to force a conveyor shutdown, Earth would have to undergo a 4 to 5 degree C warming."[8] Based on this statement and Item 4 (p. 61) of the 2001 IPCC report, comment on the possibility of a conveyor shutdown in the North Atlantic before 2100.

**Question 6-4:** Sulfur dioxide ($SO_2$) is a pollutant produced mainly by coal and oil combustion. It is a toxic chemical in high concentrations and is a major contributor to acid precipitation. Pollution from Chinese coal plants puts vast quantities of $SO_2$ into the upper atmosphere, where it may help to explain why global temperatures have not been as high over the past half-decade as models forecast. ($SO_2$ in the upper atmosphere disperses incoming sunlight and thus may mediate atmospheric temperature increase.) Would you favor relaxing pollution reduction regulations to help reduce global warming, as some have proposed? Discuss.

---

[8] Broecker, W.S. 1999. What if the conveyor were to shut down? *GSA TODAY, 9* (1): 2–7.

# SEA-LEVEL CHANGES AND GLOBAL WARMING

Most scientists think that global warming will result in rising sea level, as polar regions warm and polar ice melts. Any increase in sea level due to global warming would result from (1) the melting of ice in glaciers and polar regions and (2) the thermal expansion of seawater. Whereas the scale of the former is difficult to assess, the impact of sea-level rise from the thermal expansion of the water is more easily evaluated arithmetically.

## Thermal Expansion of Seawater

The *coefficient of thermal expansion (CTE)* of seawater is approximately 0.00019 per degree Celsius. This simply means that, given a volume of seawater, that volume will expand by this fraction as a result of heating the water, per degree. Now since ocean basins are constrained by their bottoms and "sides," the only way to go is up. If a volume of seawater occupied 1 cubic meter of water (1000 L, 264.20 gal), after warming by 1°C, it would expand to 1.00019 m$^3$ (1000.19 L, 264.25 gal). Note that this also could result in extensive coastal flooding.

To calculate how much a temperature increase would increase sea level, simply multiply the average ocean depth (in cm) by the coefficient of thermal expansion by the number of degrees of temperature rise. Note that we put a 1 before the CTE value to obtain the height of sea level after the warming. If you simply multiply the average depth by the CTE, you will get only the number of centimeters that sea level will rise.

**Question 6-5:** How much would each 1°C increase in seawater temperature cause sea level to rise? Express your answer in cm (30 cm = 1 ft). Recall the average depth is 3,800 m (3.8 km).

Whereas thermal expansion acts upon water already in the basin, melting ice represents water added to the present ocean volume. The melting of ice that is currently perched upon terrestrial land, as in the ice sheets of Greenland, Iceland, and the Antarctic, has the potential to raise sea level by about 80 meters (260 ft). Ice that is already floating in the ocean water (Arctic ice, Antarctic ice shelves, and icebergs) may melt but will not contribute to sea-level rise. The mass of water contained in these features already displaces approximately its equivalent water volume.

A recent EPA report concluded, "The total cost for a one meter rise would be $270–475 billion (in the U.S.), ignoring future development. We estimate that if no measures are taken to hold back the sea, a one meter rise in sea level would inundate 14,000 square miles, with wet and dry land each accounting for about half the loss. The 1500 square kilometers (600–700 square miles) of densely developed coastal lowlands could be protected for approximately one to two thousand dollars per year for a typical coastal lot. Given high coastal property values, holding back the sea would probably be cost-effective."[9]

---

[9] www.epa.gov/climatechange/effects/coastal/slrmaps_cost_of_holding.html.

**Question 6-6:** Discuss whether or not you agree that "holding back the sea" would be the best solution to rising sea level. What other options exist?

## Which Areas Will Be Affected?

Sea-level rise will increase flood risks in areas already at or under sea level, like New Orleans, Louisiana, in the United States (already the site of disastrous 2005 and 2012 floods) and coastal Holland in Europe. In Bangladesh, about 17 million people live less than 1 meter above sea level. In Southeast Asia, a number of large cities, including Bangkok, Mumbai, Calcutta, Dhaka, and Manila (each with populations greater than 5 million) are located on coastal lowlands or on river deltas. Particularly sensitive areas in the United States include the states of Florida and Louisiana, coastal cities, and inland cities bordering estuaries.

Moreover, low-lying islands in the Pacific (Marshall, Kiribati, Tuvalu, Tonga, Line, Micronesia, Cook), Atlantic (Antigua, Nevis), and Indian Oceans (Maldives) will be greatly impacted. For example, in the Maldives most of the land is less than 1 meter (3.05 ft) above sea level. A seawall recently built to surround the 450-acre capital atoll of Malè cost the equivalent of twenty years of the Maldive's gross domestic product, according to UN reports.

**Question 6-7:** In 2009, the U.S. Climate Change Research Program published "Global Climate Change Impacts in the U S."[10] Read and summarize their "Key Findings" section. Which cities are at greatest risk from climate change?

**Question 6-8:** Summarize the major points of this Issue.

**Question 6-9:** Discuss the issue of global climate change from the standpoint of sustainability.

---

[10] http://www.globalchange.gov/publications/reports/scientific-assessments/us-impacts/key-findings.

## FOR FURTHER THOUGHT

**Question 6-10:** Most proposals to reduce the impact of climate change focus on (1) developing more nonpolluting renewable energy sources, (2) adding more nuclear power plants, (3) reducing energy use by conservation, (4) reforestation to "soak up" more $CO_2$, or (5) removing carbon in fossil fuels before combustion and "sequestering" that carbon in underground reservoirs like aquifers. Some have suggested that societies simply learn to live with climate change.

For now, let us assume that reducing emissions of greenhouse gases becomes a top global priority. After researching this issue, rank the list of options described above in order of decreasing practicality and effectiveness. If you find other options, feel free to include them as well. Good places to find information are the websites of the United Nations Environment Programme, the U.S. Dept of Energy, the U.S. Environmental Protection Agency (www.epa.gov), the National Oceanic and Atmospheric Administration (www.noaa.gov), and environmental organizations such as the Sierra Club and Environmental Defense. Many "think tanks" such as the Rand Corp. (www.rand.org) offer useful reports and commentary. Technology already exists to address the impacts of climate change. The decision to address climate change will be a political one.

**Question 6-11:** There are mounting concerns that rising atmospheric $CO_2$ concentrations will cause changes in the ocean's carbonate chemistry system, and that those changes will affect some of the most fundamental biological and geochemical processes of the sea. Go to http://www.pmel.noaa.gov/co2/story/Ocean+Acidification, study the issue, and discuss the impacts of acidifying the oceans.

# Issue 7

## OIL AND NATURAL GAS

### KEY QUESTIONS

- What are proven reserves of oil and gas?
- How fast are we consuming oil and gas?
- Where are global reserves of oil and gas?
- What is hydrofracturing, and why is it controversial?
- What are the environmental effects of oil and gas use?
- What are the issues surrounding drilling for oil in the ANWR?
- What is *peak oil,* and has it been reached yet?
- How is oil and gas use related to sustainability?

### BACKGROUND

Oil is the energy basis of modern industrial society. For over fifty years after the first successful well was drilled in the mid-nineteenth century, oil use was insignificant. Although he thought his vehicles would be powered by peanut oil, Henry Ford changed all that with his mass-produced automobiles. World War I generated immense demand for gasoline-powered vehicles. During the period 1919 to 1949, oil gradually overtook coal as the most important fuel source in the United States. Today, oil provides 40% and natural gas provides more than 23% of U.S. energy. Oil provides virtually all of our transportation fuel.

By 2012, global oil demand was around 88 million barrels (1 barrel = 42 U.S. gal) per day, with the United States responsible for more than 18 million barrels per day.[1]

**Question 7-1:** What percent of global oil demand is accounted for by the United States?

**Question 7-2:** What percent of the world's 7.05 billion population is represented by the 310 million of the U.S.?

---

[1] U.S. Energy Information Administration, www.eia.gov.

Because of growing demand for oil from Asian economies, especially India and China, as well of political turmoil in oil producing countries, mainly the Middle East, oil prices have risen sharply. Chinese demand alone accounted for 9.4 million barrels a day (mb/d), more than half that of the U.S. As recently as 1994, Chinese demand was only around 3 mb/d. Indeed, since 2008, demand in Brazil, Russia, China, and India has increased by 3.7 mb/d, while demand in the U.S. and Eurozone countries has fallen by only 1.5 mb/d.[2]

## ORIGIN, DISTRIBUTION, AND EXTRACTION OF OIL

Oil is not a renewable resource. The oil fields of today originated tens of millions of years ago when organic remains were buried within sediments in the absence of oxygen. The organic matter was subjected to a critical combination of pressure (caused by deep burial) and increased temperatures. Over millions of years, the organic matter reorganized into more volatile organic molecules that we call *crude oil* or *petroleum,* usually accompanied by *natural gas.* To be extracted and used, petroleum must migrate upward, out of its deeply buried *source bed,* and into rock strata or a favorable geological structure that will trap the oil and prevent its escape (Figure 7-1).

Sometimes, pressure forces the oil all the way to the surface, where it forms *seeps,* which were known to Native Americans. The nineteenth century's oil discoveries came when "wildcatters" (oilmen who drilled speculative wells) simply drilled holes into the rocks underlying oil seeps.

Once extracted from underground reservoirs, oil is processed to remove other fluids, such as water, hydrogen sulfide, and natural gas, and then it is sent to refineries. There the oil is heated in the absence of oxygen to break or "crack" the molecules into lighter forms, which emerge as products such as gasoline, diesel fuel, or asphalt. Much, but not all, associated sulfur is usually removed at refineries as well. The sulfur that is left in motor fuels when burned forms oxides of sulfur (SOx), a toxic air contaminant, which cannot be removed by the present generation of catalytic converters. The oil may then be shipped by pipeline or oil tanker to destinations around the world.

## HOW MUCH OIL IS LEFT?

Calculating the amount of oil present in all known world oil fields that can be extracted at a profit using present technology yields a value called *proven oil reserves* (Table 7-1).

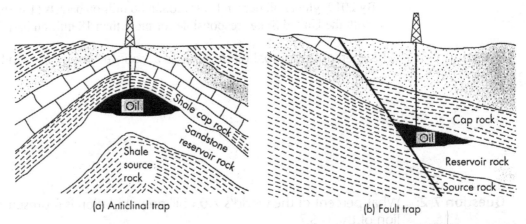

**FIGURE 7-1** Typical oil and gas "traps." Oil, being lighter than water, floats to the top of the reservoirs. The oil is usually associated with natural gas as well. (Keller, E.A. 2000. *Environmental Geology* 8th ed., Fig. 15.10, page 411. Prentice Hall, Upper Saddle River, NJ. Courtesy of E. A. Keller/Pearson Education.)

[2] *The Economist,* 6/23/12 p. 73.

**TABLE 7-1** ■ Proven Oil Reserves (billion barrels) by Country, 2011[1]

| Country | Reserves |
|---|---|
| 1. Saudi Arabia | 262 |
| 2. Venezuela | 211 |
| 3. Canada* | 175 |
| 4. Iran | 137 |
| 5. Iraq | 115 |
| 6. Kuwait | 104 |
| 7. U.A.E. | 98 |
| 8. Russia | 60 |
| 9. Libya | 46 |
| 10. Nigeria | 37 |
| 11. Kazakhstan | 30 |
| 12. Qatar | 25 |
| 13. U.S.A. | 20 |

*Canada has over 130 billion barrels of waxy petroleum solids in sedimentary deposits called "tar sands." Although they could be, and are, extensively mined, they are not conventional deposits.
[1]Source: US CIA.

Organizations such as the U.S. Department of Energy and the International Energy Agency collect and publish such information. World proven oil reserves are less than the actual oil in place. Variable amounts of oil in underground reservoirs always remain in the rock. However, new technology may increase oil recovery, or the price may go up making it economic to spend more money to get more oil out. Moreover, companies frequently underestimate the amount of oil in a field or may not publicize the actual amount for competitive reasons. For example, in 1970 BP estimated its Forties North Sea oil field contained 1.8 billion barrels of proven reserves. However, by 2012, BP, and new owner Apache, had produced 4.0 billion barrels from the field!

If oil price rises, marginal fields—those not economic to develop at current prices—may become profitable and may be added to reserves. Chevron Oil Corporation estimated in 1991 that 6,700 billion barrels of oil could be ultimately recovered, assuming *a price of $60 per barrel in 1990 dollars.*[3]

**Question 7-1:** According to the Department of Labor, the consumer price index (CPI) is a measure of the representative cost of a "basket" of goods and services. It stood at 127.4 in 1990, and had reached 230 by mid-2012.[4] Therefore, by what percent had the CPI increased between 1990 and 2012?

[3] Holyoak, A.R. Written communication.
[4] U.S. Department of Labor Bureau of Labor Statistics (http://www.bls.gov/bls/).

**Question 7-2:**   To determine the 2012 price that would be equivalent to a 1990 price of $60 per barrel, first multiply the 1990 minimum price by the percentage you just calculated. What would be the 2012 price needed to "guarantee" 6,700 million barrels of ultimately recoverable oil?

**Question 7-3:**   For a world price of $100+ per barrel to stimulate increased oil production, the price would have to be maintained long enough for companies to be able to make investment decisions on it, since producing oil from a new discovery can take ten years. What do you think would happen to oil exploration and ultimately production if the world price were to fall precipitously?

# TECHNOLOGY ENHANCES OIL DISCOVERIES AND PRODUCTION

Over the past several decades, technology has helped discover new oil and gas fields. It has also increased ultimate recovery of the oil and gas in place, both in known fields and new discoveries. Companies use seismic data and powerful computers to create 3-D images of deeply buried geologic structures more than 5 kilometers below the surface. And recent advances in drilling techniques have lowered production costs and increased oil and gas yields from known fields. Using new techniques, companies have increased the ratio of discoveries to "dry holes." Dry holes are exploratory wells that didn't hit oil and can cost up to $15 million each!

For the past twenty-five years, with only two exceptions, estimates of ultimately recoverable oil have ranged from around 1,800 billion barrels to around 2,400 billion barrels. We will discuss a new method of natural gas exploration and production, hydrofracturing, below.

## Where Is the World's Oil?

As you can see from Table 7-1, the Persian Gulf region contains more than half of the world's proven oil reserves. It similarly produces half the world's oil. The Gulf region is such an extraordinary source of oil because it has huge oil *fields* and extremely low costs of exploration and production. For example, it can cost as little as a few dollars per 42-gallon barrel in the Persian Gulf to produce oil. In new oil fields of North America and the North Sea of Europe, costs can range from $20 to $60 a barrel.

**Question 7-4:**   What is today's world price for oil? Check the web, a daily newspaper, or a business network.

We can determine how long world proven reserves (1,300 billion barrels at the end of 2011) will last by simply dividing the total oil by the annual production—approximately 32 billion barrels per year.[5] Note that when we do this, the resultant number is years since "barrels" cancel out.

---

[5]   U.S. Central Intelligence Agency (CIA), www.cia.gov.

This number represents the total number of years from 2011 that oil could be produced at current rates before it is all gone, *assuming* constant production *and* demand. In reality, the depletion of a resource like oil does not follow such a simple pattern. Rather, oil production will gradually decline over a long period of time.

**Question 7-5:** Assuming no new discoveries, in what year, starting from 2011, would the world run out of oil?

However, the rate of consumption is increasing each year, and new discoveries are being made (Table 7-2).

Consult Table 7-2. Describe the change in global demand from 1986 to 2012.

**TABLE 7-2** ■ World Oil Demand[2]

| Year | Consumption (million barrels per day) |
|------|---------------------------------------|
| 1986 | 61.8 |
| 1987 | 63 |
| 1988 | 64.8 |
| 1989 | 65.9 |
| 1990 | 66 |
| 1991 | 66.6 |
| 1992 | 66.7 |
| 1993 | 67 |
| 1994 | 68 |
| 1995 | 70 |
| 1996 | 71.7 |
| 1997 | 73.7 |
| 1998 | 73.6 |
| 1999 | 74.7 |
| 2000 | 75.8 |
| 2001 | 76.4 |
| 2002 | 77.8 |
| 2003 | 79.3 |
| 2004 | 82.6 |
| 2005 | 83.7 |
| 2006 | 85.1 |
| 2007 | 85.8 |
| 2008 | 85.4 |
| 2009 | 85.6 |
| 2010 | 87.0 |
| 2011 | 88.0 |
| 2012 (est) | 88.7 |

[2] U.S. Energy Information Administration. International oil data for crude oil production, available at http://www.eia.doe.gov.

**Question 7-6:**   What was the average annual growth rate in oil consumption from 1986 to 2012? (The most accurate way to calculate the average rate of increase is to use the formula $r = (1/t)\ln(N/N_0)$; see "Using Math in Environmental Issues," pages 6–8).

It is important to note how a tiny rate of increase can lead to such a change in oil consumption over a long period of time. The concept of doubling time, $t = 70/r$ introduced earlier, can be applied to illustrate growth of consumption.

**Question 7-7:**   What is the doubling time for the average increase in demand you just calculated?

**Question 7-8:**   To estimate the total amount of oil that was consumed over a given period, first convert the beginning and ending rates of consumption (say, 80 million barrels a day) to barrels per year. Then add the beginning and ending consumption rates together and divide by 2. Finally, multiply this number by the number of years. How much oil was consumed between 1986 and 2011?

**Question 7-9:**   EPA reports that 0.43 tonnes of $CO_2$ is emitted per barrel of oil burned. How much $CO_2$ was emitted by burning this much oil between 1986 and 2011?

Author and environmental provocateur Garrett Hardin, in addressing the issue of energy supply, said, "Whenever it is thought to be impossible to limit the growth of either population or desire, *it is impossible to solve a shortage by increasing the supply*" [emphasis his].[6]

**Question 7-10:**   Explain why you agree or disagree with this statement. Can we ever accommodate the world's increasing demand for oil? (If you have some background in economics, cite reasons why an economist might disagree with Hardin's quotation).

[6] Hardin, G. November/December 1996. Letter to Editor. *Worldwatch Magazine,*

## ANWR's Oil Reserves

The American Petroleum Institute, among others, would like to open the Arctic National Wildlife Refuge (ANWR) on Alaska's North Slope to oil exploration.[7] The U.S. Geological Survey (USGS) estimates between 5.7 and 16 billion barrels ultimately might be produced over several decades.[8] Environmental scientists are concerned about (1) the impact of oil exploration on the tundra environment and (2) the impact of oil exploration and production on caribou and other migrating animals.

**Question 7-11:** The mean estimate by the USGS of recoverable oil in ANWR is about 10.3 billion barrels without regard to price. Based on the mid-2012 U.S. annual consumption rate of about 7 billion barrels of oil per year, how many months would these new fields last?

**Question 7-12:** Did we phrase Question 7-12 fairly and accurately? Why or why not? Cite evidence for your position.

## Impacts of Oil Refining

Oil refineries are one of the top sources of industrial air pollution in the United States, and a dangerous place to work. For example, in 2010 four workers were killed and three severely injured in an explosion at a refinery in Washington State. And in 2005 a fire and explosion at a Texas refinery killed 15 workers and injured more than 170. Moreover, refineries are one of the largest stationary sources of volatile organic compounds, the primary component of urban smog. They are the fourth largest industrial source of toxic emissions and the single largest source of benzene emissions, which are carcinogenic.

**Question 7-13:** Research this issue and explain why oil refineries are so dangerous.

---

[7] American Petroleum Institute (www.api.org).
[8] U.S. Geological Survey, www.usgs.gov.

**Question 7-14:** Most politicians express horror at the prospect of higher oil prices. Identify advantages and disadvantages of higher prices for oil and refined products like gasoline and diesel fuel.

China became a net oil importer in 1995, and the Chinese demand for petroleum has increased significantly since 2001. Chinese oil demand was 9 million barrels a day in 2010 and was forecast by U.S. EIA to increase to 19 million barrels a day by 2020.

**Question 7-15:** What are the implications for the world oil price if Chinese demand sharply increases?

# NATURAL GAS

Methane, $CH_4$, is the main constituent—about 75 percent—of natural gas, but natural gas usually also contains ethane ($C_2H_6$), propane ($C_3H_8$), and butane ($C_4H_{10}$). It is one of the planet's most important commodities. It presently meets more than 20 percent of world energy needs, and nearly a quarter of the energy needs of the United States. Since burning gas produces no SOx pollution, no ash or toxic emissions like heavy metals, and half the $CO_2$ of coal, natural gas is poised to be the fastest-growing fossil fuel source in the twenty-first century.

Natural gas can be produced in any geologic environment in which organic matter is decomposed by microorganisms in the absence of oxygen, with or without heat and pressure.

## Deposits of Natural Gas

Natural gas deposits fall into four categories:

1. Natural gas is usually found with petroleum in reservoir rocks.
2. Important reserves of gas are also found in rocks with little or no oil, most recently in black shales.
3. Natural gas originates with coal and is typically found with coal deposits.
4. Methane can be formed by the action of certain bacteria in oxygen-free environments, such as waterlogged soils in permafrost regions and in deep marine sediment.

Methane deposits in deep marine sediments are called *methane hydrates*. Recent research on methane hydrates in the world's oceans point to hydrates as a potential new category of natural gas resources, though none is produced at present.

## Transporting Natural Gas

Natural gas may be transported in two ways: in pipelines (the Chinese built the first ones of bamboo in the sixth century BCE) as a pressurized gas, or as a super-cooled liquid (LNG, liquified natural gas) using specially constructed tankers.

## Global Gas Reserves

The U.S. EIA estimated global proven reserves of natural gas at 6,500 trillion cubic feet (TCF) in 2012. In 2011, the world used more than 113 TCF of gas, and this quantity was projected to increase at 1.6% per year through 2035. In the United States, natural gas consumption averaged around 22 TCF during 2005–2011. Reserves of natural gas are shown in Table 7-3. This does not include coalbed methane. The volume of gas in coal seams is more uncertain. What is certain is that methane seeping from coal mines is a major greenhouse gas, as well as a valuable potential resource. The presence of natural gas along with coal poses the biggest hazard to miners working underground, since it is under pressure, is odorless and colorless, and easily ignited. The Chinese government acknowledges that several thousand coal miners die yearly, and labor activists argue the figure is much higher. Most of these deaths were caused by methane explosions.

Even though burning methane produces $CO_2$, the primary greenhouse gas, methane is a far more powerful greenhouse gas than $CO_2$. Extraction of methane in deep marine sediments is not presently economically viable, although the amount of methane in such deposits is enormous.

**TABLE 7-3 ■** World Natural Gas Reserves by Country 2005 and 2011 (trillion cubic feet)

| Country | Reserves, 2005 | Reserves, 2011 |
|---|---|---|
| World | 6,040 | 6,500 |
| Russia | 1,680 | 1,680 |
| Iran | 940 | 1,050 |
| Qatar | 910 | 900 |
| Saudi Arabia | 235 | 280 |
| United States | 189 | 277 |
| Turkmenistan | 74 | 270 |
| United Arab Emirates | 212 | 230 |
| Nigeria | 176 | 190 |
| Venezuela | 151 | 180 |
| Algeria | 161 | 160 |
| Iraq | 110 | 115 |
| Australia | — | 110 |
| Indonesia | 90 | 105 |
| Kazakhstan | 66 | 85 |
| Malaysia | 29 | 85 |
| Egypt | 57 | 75 |
| Norway | 75 | 72 |
| Uzbekistan | 71 | 65 |
| Netherlands | 65 | — |
| Canada | 62 | 60 |
| Ukraine | 40 | 40 |

Source: CIA World Factbook (https://www.cia.gov/library/publications/the-world-factbook/rankorder/2179rank.html )

## Hydraulic Fracturing

Hydraulic fracturing ("fracking") permits gas production from rocks containing gas but with very low permeabilities, such that the gas cannot be extracted using conventional methods. Fracking injects large volumes of fluids and "proppants"(small particles of solids) at high pressure, fracturing the rock, which allows gas to escape. The fluids inject the proppants into the fractures to keep them from sealing shut when production begins. The fracking fluid is typically water-based and contains such chemicals as bactericides, buffers, fluid-loss additives, and surfactants (essentially, detergents), to make the fracturing efficient and prevent damage to the formation.

There are two problems with fracking: first, companies usually, for proprietary reasons, are reluctant to disclose the precise content of the fracking fluid, and second, large amounts of water are used, whose content is modified by mixing with water from deeply buried rocks, as well as the fracking fluid itself. Potential environmental impacts include contaminated wells, and local water supplies may be threatened. One of the nation's greatest reservoirs of "tight" shale gas, the Marcellus Shale, underlies much of the eastern United States, but production from the formation may be severely restricted if means to prevent water contamination are not perfected.[9] Indeed, half the new natural gas wells in the U.S. involve fracking. Globally, EIA estimates shale ("tight") gas reserves to exceed 60 trillion cubic feet.

**Question 7-16:**   Global gas consumption is projected by the U.S. Energy Information Administration to grow at 1.6% percent per year. Use the doubling time formula to project how long at this rate it will take for consumption of natural gas to double from 2011's 113 TCF per year.

**Question 7-17:**   From the perspective of global sustainability, what role should natural gas play in the world's energy future? Why?

**Question 7-18:**   Summarize the important points of this Issue.

# FOR FURTHER THOUGHT

## Natural Gas from Prudhoe Bay

The Prudhoe Bay oil field in Alaska's North Slope has large reserves of natural gas, estimated by field operator BP at 35 TCF, but no way to get the gas to market. Gas cannot be transported using the existing oil pipeline, which cost over $7 billion to build in the 1970s.

---

[9] West Virginia Rivers Coalition www.wvrivers.org/articles/Marcellus%20Report%202010.pdf.

**Question 7-19:** By what two ways could the Prudhoe Bay gas be transported to a potential market?

**Question 7-20:** As shale gas has become more common since 2007, the price of natural gas has fallen by two thirds. How would this affect the economics of a Prudhoe Bay gas pipeline?

**Question 7-21:** Do you think most Americans believe that our oil and gas reserves are being depleted at a rate that will exhaust them in the twenty-first century?

**Question 7-22:** Ecological economists contend that were externalities to be factored into the cost of fossil energy, the price would increase substantially. Determine what these externalities might be. You could consult an article by Hubbard [10] and a paper by Roodman.[11] Also read a CEERT report.[12] A web search for "energy externalities" will yield much useful information as well. List the sources of these hidden costs, and discuss whether consumers should pay them, and, if so, what effect they might have on energy use.

**Question 7-23:** Consider the issue of "peak oil." The production history of a conventional oil field displays an inverted "U" shape, as shown in Figure 7-2. If one extrapolates total global production from the sum of individual fields, one arrives at a concept called peak oil. In other words, at some time in the history of global oil production, production will "peak" and then begin to decline. U.S. domestic production "peaked" in around 1970 and has been in decline since. Research the concept of peak oil for global oil reserves. When is global production forecast to peak? Assuming the concept of peak oil is accurate, discuss the extent to which oil and gas will contribute to sustainable societies.

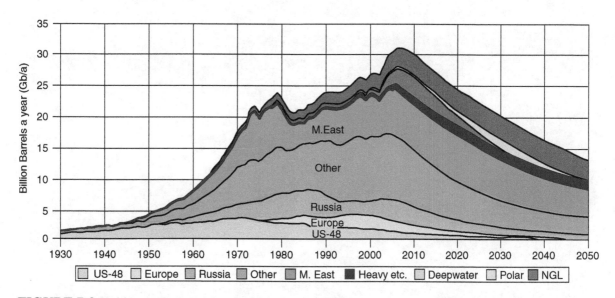

**FIGURE 7-2** Hubbert's Curve, named after M. King Hubbert, whose model depicts the shape of the oil production curve.

[10] Hubbard, H.M. April 1991. The real cost of energy. *Scientific American*, 264: 36–42.

[11] Roodman, D.M. 1996. Paying the piper: Subsidies, politics, and the Environment. Worldwatch Institute paper 133. Worldwatch Institute, Washington, DC.

[12] http://www.ceert.org.

**Question 7-24:** Many politicians from corn-growing states like Iowa have asserted that ethanol use can free the United States from dependence on imported oil. Research the pros and cons of ethanol.

**Question 7-25:** Research the "Keystone XL Pipeline" and "tar sands" (also known as *oil sands* and *bitumen* sands) and summarize the main issues surrounding the controversy. Explain what Dr. James Hansen, head of NASA's Goddard Institute for Space Studies and the U.S. government's chief climate scientist, meant when he stated this about the proposed Keystone XL pipeline: ". . . if the tar sands are thrown into the mix, it is essentially game over . . ." (for the climate).[13]

**Question 7-26:** Dr. Hansen also wrote this about tar sands[14]: "The environmental impacts of tar sands development include: irreversible effects on biodiversity and the natural environment, reduced water quality, destruction of fragile pristine Boreal Forest and associated wetlands, aquatic and watershed mismanagement, habitat fragmentation, habitat loss, disruption to life cycles of endemic wildlife particularly bird and Caribou migration, fish deformities and negative impacts on the human health in downstream communities." In light of what you have learned about the proposed Keystone XL pipeline and tar sands, explain whether you would or would not support construction of the pipeline and further development of Canadian tar sands.

---

[13] Available at: http://www.countercurrents.org/hansen040611.htm.
[14] Ibid.

# Issue 8

## COAL

### KEY QUESTIONS

- What are trends in global coal use?
- What emissions are produced when coal is burned?
- Does Chinese coal use affect the "trade deficit" with the United States?
- What is the impact of coal use on the global environment?
- What is the impact of Chinese coal use on its own citizens?
- Are there ways to use coal without harming the environment?

### COAL, AIR POLLUTION, AND THE U.S.-CHINA TRADE DEFICIT

You are probably aware that the United States runs a large "trade deficit" with the People's Republic of China—meaning China sells far more to the United States than it buys. During the past two decades, the United States has lost many thousands of manufacturing jobs to China (and other developing countries). The reason usually given by business leaders is a cost advantage for Chinese goods. One such cost is energy—both the price paid by manufacturers and the real cost to society in the form of *externalities,* such as air and water pollution and deaths and injuries in the energy industry itself.

In 2000, the world burned about 4,000 million tonnes of coal. In 2011, it burned nearly 8,000 million tonnes. This increase in coal use was almost entirely accounted for by the People's Republic of China (PRC), and, to a much lesser extent, India. China used nearly as much coal in 2011—3.8 billion tonnes—as the rest of the world *combined.*[1] And the Chinese are increasing coal use at the rate of about 14 percent per year.

**Question 8-1:** What is the doubling time for a 14 percent annual rate of increase? (Recall that the doubling time formula is t = 70/r, see "Using Math in Environmental Issues," pages 6–8).

---

[1] US CIA, World Factbook, www.cia.gov.

# SOME EXTERNALITIES OF COAL USE IN CHINA[2]

According to the World Bank, about 180,000 people die each year in China from exposure to "normal" air pollution levels in cities, caused mainly by industrial coal burning.[3]

■ China produces more than one-sixth of the world's toxic SOx pollution.
■ Particulates from Chinese coal plants and industries make a significant and rising contribution to particulate levels in cities of western North America, making it more expensive to meet increasingly stringent air pollution standards in California and elsewhere. In fact, the Chinese emit so much SOx that it has been implicated in a short-term cooling of the Earth over the past decade, contrary to the predictions of climate change models, which did not include the massive air pollution.
■ China emits more mercury pollution than any other country in the world.
■ In 2005 more than 6,000 miners were killed in Chinese mine accidents, according to government statistics. Labor activists claimed the figure to be closer to 10,000.

**Question 8-2:**  What was the death rate in Chinese mines in 2005, using official figures, in deaths per million tons mined? China mined about 1.5 billion tonnes of coal in 2005.

**Question 8-3:**  Is it likely that the costs of coal use not included in the price affects the competitiveness of Chinese manufactured goods sold in the United States? What evidence did you use to formulate your answer? What additional evidence would you like to have to better answer this question?

# ORIGIN AND NATURE OF COAL

Coal is a fossil fuel composed essentially of compressed altered plant remains. Under circumstances which have been replicated in many places on the planet over the past 350 to 400 million years, plant tissue has been buried under anoxic (oxygen-free) conditions, allowing plant tissue to be incompletely decomposed by anaerobic bacteria. Economically viable coal deposits form when remains of entire forests are buried and preserved by later sediments. Over millennia, such buried plant tissue may be converted through a series of stages involving microbial activity, pressure, and heat into *peat, lignite, sub-bituminous,* and *bituminous* coal, or the rarest coal, *anthracite,* the form of coal with the highest amount of energy per kilogram.

---

[2] From several sources, including *Pollution from Chinese Coal Casts Shadow around Globe,* by Keith Bradsher and David Barboza. *New York Times,* June 11, 2006, and *The Guardian,* http://www.guardian.co.uk/environment/gallery/2010/sep/16/pollution-coal-ash-china.
[3] www.worldbank.org.

Coal is composed predominantly of the element carbon (C) along with hydrogen (H) and nitrogen (N). Carbon content of coal ranges from 45 percent up to as much as 98 percent in rare anthracite. Most coal used to produce electricity contains between 50 and 70 percent carbon. Burning coal, therefore, produces large amounts of $CO_2$, the main agent of global climate change.

**Question 8-4:** "Burning" 1 kilogram of C produces about 3.7 kilograms of $CO_2$. Assume typical coal contains 60 percent C. How much $CO_2$ is released per tonne of coal burned?

**Question 8-5:** The United States produced about 1,050 million tons (= 1,000 million tonnes) of coal in 2010. How much $CO_2$ was released in 2010 when that was burned?

**Question 8-6:** About 8,000 million tonnes of coal were burned globally in 2011. How much $CO_2$ was produced, assuming 60 percent C content?

**Question 8-7:** According to the Energy Information Administration, the world is expected to consume a minimum of 8,000 million tonnes of coal each year up to 2030, with China accounting for half that. How much $CO_2$ would be produced by coal burning in 2030?

**Question 8-8:** How can global $CO_2$ emissions be controlled, in your view, if these forecasts prove accurate? Discuss the implications if they are not controlled (see Issue 6, for example).

## COAL BED METHANE

When plant matter is converted into coal, methane can also be formed, the main ingredient in natural gas (Issue 7). The methane may be trapped in the sedimentary rock containing the coal. Some of this trapped gas is released when coal is mined, which can lead to dangerous explosions. These explosions are the reason for most miners' deaths in underground mines. Large mines may release up to 1 million cubic feet per day of methane. This can be a significant source of atmospheric methane, a powerful greenhouse gas (see Issues 5 and 6). It is also a potentially valuable energy resource.[4]

## OXIDATION OF PYRITE AND WATER POLLUTION

One of the most widespread and polluting reactions is that of the oxidation of the mineral pyrite, $FeS_2$, also called "fool's gold." Pyrite is found in virtually all coals, as well as in the fossil soils immediately below the coal seams. When the coal is stripped away, the pyrite-bearing sediment is exposed to oxidation and hydration, producing sulfuric acid. The shale above coal seams also contains a lot of pyrite, and shales are often left on "spoil heaps" to oxidize. Strip-mining coal may thus lead to extensive acidification of surface water.[5]

## AIR POLLUTION FROM COAL COMBUSTION

In addition to $CO_2$, coal burning can produce (1) high levels of toxic sulfur oxide gases (SOx), (2) appreciable quantities of toxic heavy metals such as mercury, uranium, and chromium, and (3) oxides of nitrogen (NOx). SOx and NOx can contribute to smog and acid precipitation, which can damage buildings and cause pulmonary problems. Heavy metals can build up in plant tissue and the tissue of aquatic organisms.

Coal burning is a major contributor of mercury (Hg) to the environment. Mercury levels in freshwater and marine fish in many states are high enough to trigger health warnings. Levels of methylmercury, the most toxic form, that might not harm an adult can damage a child's developing brain, lead to deficiencies in IQ, cause attention deficit, and impair motor function. Fetuses and infants are especially vulnerable. Figure 8-1a is a map of the United States showing fish advisories due to methylmercury contamination. Atmospheric mercury concentration is shown in Figure 8-1b. About two-thirds of SOx and about one-fourth of NOx is from coal-burning power plants.[6]

**Question 8-9:** What regions of the United States are most susceptible to atmospheric mercury contamination?

An "average" 500-megawatt coal-burning power plant each year emits pollutants as shown in Table 8-1.[7]

---

[4] See for example U S Geological Survey, pubs.usgs.gov/fs/fs123-00/fs123-00.pdf.
[5] See Environmental Protection Agency, htttp://water.epa.gov/polwaste/nps/acid_mne.cfm.
[6] U.S. EPA.
[7] U.S. EPA. See also: http://web.mit.edu/coal/The_Future_of_Coal.pdf.

**TABLE 8-1** ■ Emissions of an average 500-megawatt coal-fired power plant (EPA).

| |
|---|
| 100 kg arsenic |
| 2 kg cadmium |
| 50 kg lead |
| 450 tonnes particulates (t) |
| 4,500 (t) sulfur oxides (SOx) |
| 9,000 (t) nitrogen oxides (NOx) |

Fish Consumption Advisories for Mercury

Explanation

- Statewide Advisory
- Other States with Mercury Advisories
- States with Coastal Advisories

**FIGURE 8-1a** Map of the United States showing fish advisories related to methylmercury contamination (USGS).

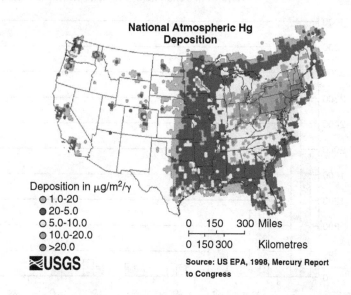

National Atmospheric Hg Deposition

Deposition in μg/m²/γ
- 1.0-20
- 20-5.0
- 5.0-10.0
- 10.0-20.0
- >20.0

**USGS**

0    150    300  Miles
0  150 300      Kilometres

Source: US EPA, 1998, Mercury Report to Congress

**FIGURE 8-1b** Modeled Atmospheric Mercury Deposition (2005, from USGS).

## COAL AS A RESOURCE

At this juncture you might be asking yourself (1) if coal mining is so hazardous to miners and produces significant levels of water pollution, (2) if $CO_2$ levels are rising appreciably due in large part to coal burning, and (3) if coal burning is responsible for much if not most of the air pollution in developing countries, why is there such a demand for coal?

In spite of these issues, coal is a critically important resource. This is due to four major factors: (1) the relative ease and low cost with which coal may be mined and transported (and these costs are low in large part because externalities are ignored), (2) the energy readily made available when coal is burned, (3) the lack, or nonenforcement, of environmental laws in many developing countries, and (4) the strides made in most industrialized countries to improve mine safety and control certain coal emissions.

## GLOBAL COAL USE

Figure 8-2 shows global coal consumption. According to oil giant BP, coal accounts for one-fourth of global energy.

**Question 8-10:** Which four countries account for most coal use worldwide?

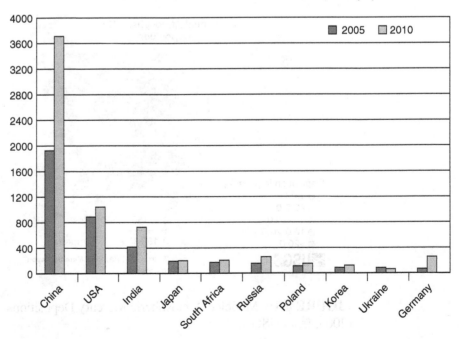

**FIGURE 8-2** Global coal consumption. (Source: World Coal Institute, www.worldcoal.org)

**Question 8-11:**   The United States' total coal production is approximately 1,100 million short tonnes (= tons) annually. How many years at this rate will domestic supplies last?

Figure 8-3 shows global recoverable coal reserves.

**Question 8-12:**   Naturalist and writer Edward Abbey once said, "We will run out of air before we run out of oil." His words are perhaps even more applicable to coal. What do you think the quote means, as applied to coal?

**Question 8-13:**   How is increasing global use of coal related to issues of sustainability?

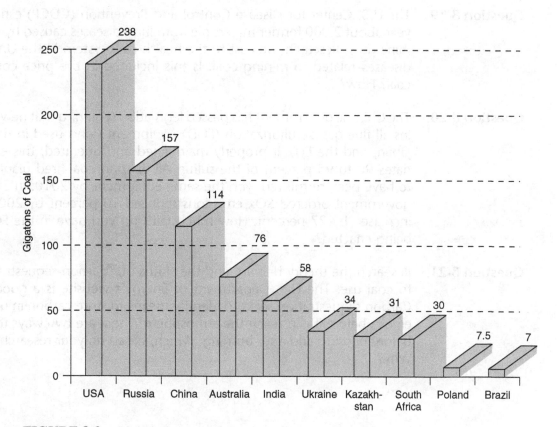

**FIGURE 8-3**   Global recoverable coal reserves. (Source: BP)

**Question 8-14:** Summarize the main points of this Issue.

## FOR FURTHER THOUGHT

**Question 8-15:** Coal companies assert that the air has gotten cleaner in the United States over an interval during which coal use has tripled. Environmental organizations decry coal's impact on the environment. For contrasting views, access the sites of coal companies such as Arch Coal, www.archcoal.com, and environmental organizations like the Sierra Club, www.sierra.org. What is your conclusion?

**Question 8-16:** The United States has among the world's safest coal mines. U.S. coal production averages close to 1,100 million short tons per year. Deaths over the period 2000–06 averaged about thirty per year. What is the death rate in U.S. mines per million tons mined?

**Question 8-17:** Research the issue of coal deaths in Chinese mines. How does the rate in Question 8-16 compare to the death rate in Chinese mines?

**Question 8-18:** Is this a contributing factor in the ability of China to "out-compete" western countries in manufacturing? Why or why not?

**Question 8-19:** The U.S. Center for Disease Control and Prevention (CDCP) estimates that each year about 2,000 former miners die from lung diseases caused by exposure to coal mine dust. Research the cost to the health care system in the United States from diseases related to mining coal. Is this included in the price consumers pay for coal? How?

**Question 8-20:** The Chinese government has passed legislation requiring that new coal-fired plants install flue gas desulfurization (FGD) equipment (long used in the United States, Japan, and the EU). If properly maintained and operated, this equipment eliminates 90 to 95 percent of the sulfur. All existing coal-fired plants in China were to have been retrofitted with the same equipment by 2010. In 2002 the Chinese government ordered SOx emissions reduced 10 percent by 2005. Instead, they increased by 27 percent. How much faith do you have in the FGD requirement being enacted?

**Question 8-21:** Research the theory behind, and the status of, "carbon sequestration" as applied to coal use. The U.S. Department of Energy's website is a good place to start. Carbon sequestration seeks to identify means to store carbon in underground reservoirs before it can reach the atmosphere. There are two ways to remove $CO_2$— before burning and after burning. Which, based on your research, is more viable? Why?

**Question 8-22:** While China is responsible for an immense and increasing amount of regional and global air pollution and $CO_2$, many scientists point out that the average American still consumes far more energy, and is responsible for ten times the carbon emissions, compared to the average Chinese citizen. How can the United States and China work together to reduce coal emissions and protect the global environment?

**Question 8-23:** Research the issues related to a surface mining practice called "mountaintop removal." This is a controversial form of surface coal mining that is changing the very face of states like West Virginia and Kentucky. This practice levels mountain tops and produces millions of tons of debris that are dumped into adjacent valleys, changing the contours and drainage patterns of thousands of square miles in the eastern United States. Summarize your findings.

**Question 8-24:** The coal industry has invested heavily in marketing "clean coal." Research the subject. What is "clean coal?" Discuss whether "clean coal" is an oxymoronic marketing ploy, the real thing, or somewhere in between. Is the mining of "clean coal" less environmentally destructive than "unclean" coal? Does "clean coal" when combusted produce fewer greenhouse gas emissions?

# BRINGING THE WORLD TO THE U.S. STANDARD OF LIVING

## KEY QUESTIONS

- What would be the environmental impact of bringing world oil consumption to U.S. levels?
- What is the relationship between U.S. oil consumption and our standard of living?
- Would bringing the world to the U.S. standard of living be compatible with principles of sustainability?

Development agencies such as the World Bank[1] have a stated goal of enabling "developing" nations to reach a standard of living comparable to western countries such as the United States. In this Issue, we will investigate what that might involve and what kind of impact it would have on the world's oil supply and demand.

## STANDARD OF LIVING ESTIMATES

For purposes of comparison, we can estimate a country's standard of living by dividing the total annual value (in dollars) of goods and services (the gross domestic product, or GDP) by a country's population to derive the per capita GDP. To do this calculation for the United States, divide the approximate 2011 GDP ($15,300 billion)[2] by the 2011 population (approximately 311 million).[3]

**Question 9-1:** What was the U.S. per capita GDP for 2011?

*Global economic product* (GEP), or global GDP, the combined value of goods and services of all the world's nations, is harder to measure, but it approximated $65 trillion in 2011,[4] according to the International Monetary Fund (IMF).

---

[1] See www.worldbank.org.
[2] Bureau of Economic Analysis, http://www.bea.gov.
[3] U.S. Census Bureau, www.census.gov.
[4] *The Economist*, May 2012.

**Question 9-2:** Given a global population of approximately 7.05 billion in 2011, calculate the global per capita GDP for 2011.

**Question 9-3:** By what factor (2×, 4×, etc.) would the per capita GDP have to be increased to U.S. per capita GDP to equal the U.S. GDP in 2011?

# USING OIL CONSUMPTION TO ESTIMATE LIVING STANDARDS

The U.S. economy is heavily dependent on oil (see Issue 7). Oil provides more than 37% of the total U.S. energy supply and virtually all of its transportation fuel. According to the Energy Information Administration (EIA) the transportation sector consumes about 28 percent of all domestic energy.[5] For 2011, global oil production was about 87.5 million barrels per day, according to the U.S. CIA, and the United States consumed about 19 million barrels per day.

**Question 9-4:** What percentage of world oil production for 2011 was consumed by the United States?

To produce and consume the goods and services that Americans think of when we imagine a high standard of living requires oil, and a lot of it. Oil, for example, is essential to American agriculture. According to Pimentel and others, "the intensive farming technologies of developed countries use massive amounts of fossil energy for fertilizers, pesticides, and irrigation, and for machines as a substitute for human labor."[6] And most of this fossil energy comes from oil. So even if we cannot directly correlate oil consumption with standard of living, our way of life, which we are exporting to the rest of the world, is critically dependent on oil.

However, standard of living is much more than some threshold level of median income or level of material consumption. Many thoughtful observers prefer to use *quality of life* as a more reasonable measure of living standard. Measuring quality of life consists at least partially of evaluating intangibles such as the following:

■ Clean drinking water and unpolluted air
■ Access to quiet and solitude, especially in densely populated cities
■ Alternatives to automobiles and clogged roads in making transportation decisions

---

[5] U.S. Energy Information Administration, http://www.eia. doe.gov/.
[6] Pimentel, D., X. Huang, A. Cordova, & M. Pimentel. 1997. Impact of population growth on food supplies and environment. *Population and Environment*, 19: 11.

- Access to quality medical care at reasonable prices
- A job that pays a "living wage"
- A safe, low-crime environment
- A relatively secure old age

Nevertheless, we can get some appreciation of what would be required to allow the world's population to enjoy a U.S. lifestyle by analyzing petroleum consumption. Therefore, we will assume that a reasonable measure of our standard of living, aspired to by much of the developing world, is our per capita oil consumption.

**Question 9-5:**  Do you think this assumption is reasonable? If not, suggest alternatives and defend your conclusion with reasoning.

**Question 9-6:**  Let's analyze the consequences of bringing the world's per capita oil consumption to U.S. levels. Using the data on U.S. oil consumption given above and a 2011 population of 311 million, calculate the U.S. per capita oil consumption in 2011 in barrels per person per year. Convert that figure to gallons (1 barrel = 42 gallons).

**Question 9-7:**  To calculate world oil consumption at U.S. levels, simply multiply the U.S. per capita consumption figure by the mid-2011 world population (7.0 billion).

**Question 9-8:**  World total petroleum production in 2011 was approximately 32 billion barrels. For the world to consume oil at the rate of the United States, world production would have to increase by what factor (2×, 3×, etc.) over 2011 levels?

**Question 9-9:** Next, let's assume that the U.S. consumption were to drop by one third, to roughly that of the European Union (EU). In this new scenario, world oil production would have to increase by what factor (50%, 2×, 3×, etc.) to bring the world to the per capita consumption levels of the EU?

**Question 9-10:** Given a 2011 world population growth rate of 1.2 percent per year, how long would it take the world's population to double? By what year would this doubling occur? (See "Using Math in Environmental Issues," pages 6–8 for a review of doubling time.)

**Question 9-11:** Proven petroleum reserves are roughly 1,380 billion barrels (see Issue 7). Compare the annual oil demand of a world population consuming at U.S. levels (the answer from Question 9-7) to proven oil reserves of 1,380 billion barrels.

**Question 9-12:** Can sustainable societies be built on oil consumption at U.S. levels? Why or why not?

**Question 9-13:** Summarize the main points of this Issue.

# FOR FURTHER THOUGHT

## Producing and Transporting Petroleum

The Gulf of Mexico well blowout that destroyed the Transocean platform in 2010 illustrated the hazards in deep-water oil exploration, which is the last great undiscovered "frontier" in the search for oil. But notwithstanding this, transporting petroleum has risks as well.

Petroleum is transported by two ways, by oil tanker or by pipeline. A substantial increase in world oil consumption will involve major increases in tanker traffic, larger tankers, newer or larger pipelines, or a combination of the above. Research the issue, then answer the following questions.

**Question 9-14:** Assuming that the increased oil transportation would be by tanker and that tanker capacity remains the same, by what factor would tanker traffic have to increase?

**Question 9-15:** Most of the world's oil available for export comes from enormous oil fields in the Middle East, specifically from Saudi Arabia, Kuwait, Iraq, Iran, and the United Arab Emirates. Evaluate the geopolitical impact of an increase in oil tanker shipments from this region in the light of the past sixty years of conflict in the region.

**Question 9-16:** Tankers leaving the Persian Gulf, where the extra oil production would be most likely to occur, must navigate the northwest Indian Ocean, which averages five tropical storms per year. It is also the site of numerous spectacular coral reef communities, and over the past decade has been ravaged by pirates operating out of the "failed state" of Somalia. Already the mouth of the Persian Gulf is the site of the world's highest density of ocean shipping, averaging more than 400 million tonnes of oil per year by 1996.[7] It has also been the site of at least four major tanker accidents. Evaluate the environmental consequences of a large increase in tanker traffic.

**Question 9-17:** Each year "normal" leaks and spills from tankers and terminals dump 250,000 barrels of oil into the Persian Gulf. By what factor would this increase, based on the increased rate of tanker shipments that you previously calculated (assuming that all of the increased oil production would be shipped by tanker from the Persian Gulf)?

**Question 9-18:** Another way to assess standard of living is by housing size. Many Americans are now purchasing large houses, often with floor area of over 370 m$^2$ (4000 ft$^2$); the average size of new homes is around 2300 ft$^2$.[8] What would be the environmental impact if everyone lived in these "McMansions," or "starter castles," as they have been called?

**Question 9-19:** Identify as many high "quality of life" intangibles (such as clean air) as you can and explain your choices.

---

[7] Allen, J.L. 1997. *Student Atlas of Environmental Issues* (Guilford, CN: Dushkin/McGraw-Hill).
[8] *Statistical Abstracts*, http://www.census.gov/.

# Issue 10

## SUSTAINABLE ENERGY:
## *IS THE ANSWER BLOWING IN THE WIND?*

## KEY QUESTIONS

- What are the main sustainable sources of energy?
- What is their potential and actual role in U.S. and global energy production?
- How is Hawaii's energy situation unique?
- Can sustainable societies be supported entirely by renewable energy?

## INTRODUCTION

Renewable energy technologies presently include those that employ wind (Figure 10-1), solar, geothermal, biomass, and hydroelectric dams to produce power. But are all of these sustainable? Dams are not a truly renewable electricity source. Sediment buildup behind the dams will eventually fill the reservoir. This gives them an effective useful life of several decades to one or more centuries. And building dams has consequences: trapped sediment behind dams on the Mississippi and its tributaries is the main reason the Louisiana coast is rapidly receding, for example.

Moreover, hydroelectric dams usually have harmful effects on rivers by preventing the migration of fish species. And the construction of dams often displaces people: as many as 2 million in the case of China's Three Gorges Dam.

**FIGURE 10-1** Wind turbines at Searsburg, VT (Source: U.S. EPA)

# ASPECTS, ADVANTAGES, AND DISADVANTAGES OF RENEWABLE ENERGY

Renewable energy sources have the following in common:

- They are replenished by natural processes.
- They cause no direct air or water pollution since no combustion is involved.
- They require no shipment of fuels nor offshore oil exploration, so environmental disasters such as the 1989 Exxon Valdez oil tanker spill in Alaska's Prince William Sound, or the 2010 Deepwater Horizon well blowout, will not occur.
- They require no storage or use of toxic materials, like radioactive fuel, so they would make a poor target for terrorist attacks.

Renewable energy sources have, however, some significant limitations:

- Using present technologies, all must produce electricity instead of a more "portable" fuel such as gasoline or diesel. And batteries or other storage technologies that would allow renewable energy to be used extensively in the transportation sector have not yet been perfected.
- Solar power depends on sunlight. Lacking a viable means to store the electricity produced during the day, there must be a supplemental energy source at night.
- Both a minimum and maximum wind speed is required for wind power. Wind turbines are usually shut down for safety reasons if the wind speed exceeds 56 miles per hour (90 km/h). Large-scale wind power generation is limited to sites where threshold persistent winds occur.

Despite these limitations, energy production from renewable sources has increased substantially in recent years. For example, in 1999, for the first time, the world installed more new wind-generating capacity than nuclear capacity. By 2012, the Pacific Northwest had so much installed wind capacity that the Bonneville Power Authority had to occasionally mandate shutting down wind turbines.

# ENERGY CONSUMPTION

Energy consumption can be measured in British thermal units (BTU), typically as units of 1 quadrillion BTU (1 quad). One BTU is approximately equal to the amount of heat necessary to raise the temperature of 454 grams (1 lb) of water by 1°F. One quad equals $10^{15}$ BTU. U.S. energy consumption increased from 66.4 to 80.2 quad BTU from 1970 to 1988 but declined from 100.2 quads in 2004 to 96 in 2009, before rising to 98 quads in 2010.[1]

**Question 10-1:** What was the percentage increase in energy consumption from 1970 to 1988?

**Question 10-2:** What was the average yearly increase in energy consumption over the period 1970–2010? (Use the formula $r = (1/t)\ln(N/N_0)$; see "Using Math in Environmental Issues," pages 6–8).

---

[1] www.eia.doe.gov/.

**Question 10-3:** What is the doubling time of energy consumption, based on your answer to Question 10-2? (Use the formula t = 70/r).

**Question 10-4:** Based on the 1970–2010 rate of increase, estimate U.S. energy use in 2020, and 2030. Use the compound interest formula future value = present value × $(e)^{rt}$. What caveats would you advise about your conclusions?

**Question 10-5:** Total electric power generation from nonhydro renewables from 1990 to 2011 is given in Figure 10-2.[2] Assess the contribution of renewables to electricity production. What are the top three sources?

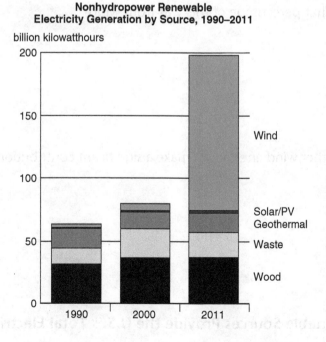

FIGURE 10-2   Nonhydroelectric renewable electricity generation, 1990–2011. (Source: U.S. Energy Information Administration, *Annual Energy Review*, Table 8.2 (October 2011), *Electric Power Monthly* (March 2012) preliminary 2011 data.)

---

[2] http://www.eia.gov/energy_in_brief/images/charts/nonhydro_renew_elec_large.jpg.

**Question 10-6:**    Study Figure 10-3. Assess the contribution of renewables to the total U.S. production of electricity. What are the top four sources of U.S. electricity? Do you see cause for optimism that renewables can make a major contribution? Explain.

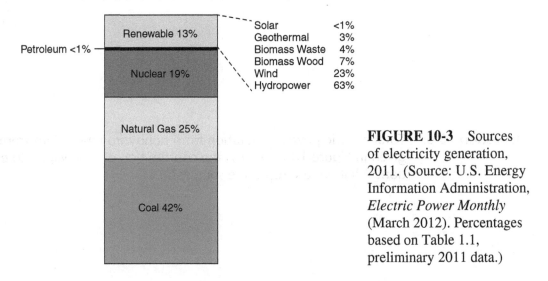

| | |
|---|---|
| Solar | <1% |
| Geothermal | 3% |
| Biomass Waste | 4% |
| Biomass Wood | 7% |
| Wind | 23% |
| Hydropower | 63% |

**FIGURE 10-3**    Sources of electricity generation, 2011. (Source: U.S. Energy Information Administration, *Electric Power Monthly* (March 2012). Percentages based on Table 1.1, preliminary 2011 data.)

**Question 10-7:**    Total electricity consumption in the U.S. in 2011 was approximately 4190 billion kWh. What percentage of this was from wind?

**Question 10-8:**    Discuss whether wind energy can make a significant contribution to U.S. electricity use in 2020.

## Can Sustainable Sources Provide the U.S.'s Total Electricity Demand?

Recall that the U.S. consumed over 4 trillion kWh of electricity in 2011. To evaluate whether sustainable sources could meet the U.S. electricity needs, you will have to consider among other things the limitations of sustainable systems. We will focus here on solar and wind power. Keep in mind that we need electricity 24 hours a day and that there are few ways to store electricity once it has been produced. And recall that electricity demand varies over the day.

**Question 10-9:** Based on your typical day, draw a chart estimating when electricity demand is greatest and least. What are some limitations of solar and wind power?

For our analysis of sustainable energy, we focus on the present and future of wind power systems.

# WIND ENERGY DEVELOPMENT IN THE U.S.

After a decade of more or less frenzied development, by early 2012 the U.S. wind industry had more than 48,600 megawatts (MW) of installed capacity, and had another 8,900 MW under construction, according to the American Wind Energy Association.[3]

**Question 10-10:** It cost about $3.5 million to install a typical 2 MW wind turbine in 2012. What was the average cost to install one kilowatt of wind energy?

**Question 10-11:** How does this compare to the cost of $700 per kilowatt to build a natural gas-fired plant?

**Question 10-12:** What advantage does wind have over gas that could offset this higher installation cost?

---

[3] American Wind Energy Association, http://www.awea.org/learnabout/.

Wind "farms" continue to be second-largest source of new power generation built in the United States, after new natural gas power plants. The United States as of 2012 ranked third in the world in wind energy, behind Germany and Spain.

According to the industry's trade association, a 1.5 MW wind turbine situated optimally should generate over 4 million kilowatt hours per year—enough to supply 400 average homes.

**Question 10-13:**   Based on information in the paragraph above, what is the annual electricity consumption of an average house in the United States?

**Question 10-14:**   According to General Electric, one of the largest turbine manufacturers in the world, 2.5 MW turbines are beginning to be installed globally. How many typical homes could be supplied by a 2.5 MW turbine, compared to a 1.5 MW turbine?

**Question 10-15:**   According to the U.S. Energy Information Administration EIA, the United States consumed about 3,800 billion kilowatt hours of electricity in 2011. How does this compare to the wind energy potential of the top five states alone in Table 10-1?

**TABLE 10-1**   ■   States with the Greatest Wind Potential
The top twenty states for wind energy potential, as measured by annual energy potential in billion kWh.

| 1 | North Dakota | 1,210 | 11 | Colorado | 481 |
|---|---|---|---|---|---|
| 2 | Texas | 1,190 | 12 | New Mexico | 435 |
| 3 | Kansas | 1,070 | 13 | Idaho | 73 |
| 4 | South Dakota | 1,030 | 14 | Michigan | 65 |
| 5 | Montana | 1,020 | 15 | New York | 62 |
| 6 | Nebraska | 868 | 16 | Illinois | 61 |
| 7 | Wyoming | 747 | 17 | California | 59 |
| 8 | Oklahoma | 725 | 18 | Wisconsin | 58 |
| 9 | Minnesota | 657 | 19 | Maine | 56 |
| 10 | Iowa | 551 | 20 | Missouri | 52 |

(*Source:* www.energyonline.com/Restructuring/news_reports/news/1011ren.html. Courtesy of Energy Online.)

**Question 10-16:** Based on these data, could wind supply the country's entire electric power demand, assuming enough turbines could be built? Why or why not? What additional information would you need to more fully answer this question?

## Some Aspects of the Economics of Wind Energy

The cost of wind power is strongly affected by three factors: average wind speed (wind power increases as the cube of its speed), interest rates (since companies typically borrow money to build the plants), and any government wind subsidy, which averaged 1.9 cents per kilowatt hour for the 2000s. The federal Wind Production Tax Credit (PTC) was first passed in 1994 and enhanced the economics of wind production through a tax incentive. It provided 1.9 cents per kilowatt hour generated for the first ten years of a wind turbine's output. The Act must be reauthorized every two years by Congress, leading to uncertainty in the future potential for wind power systems.

Even though there is no fuel cost for a wind plant, operators often pay royalties to landowners, and such plants are relatively expensive to build; thus, much of the cost is the capital required for equipment manufacturing and plant construction. This, in turn, along with the absence or presence of the PTC, means that the economics of wind is unusually sensitive to the prevailing interest rate, and politics.

Moreover, as with any form of electricity, the power must be shipped by transmission line from the generation site to the site of consumption. This limits the distance that electricity can be efficiently transported. It is impractical to produce electricity in California and ship it to New Hampshire for consumption, for example.

**Question 10-17:** The new method of hydrofracturing, or "fracking," has substantially increased natural gas reserves in the U.S. (see Issue 7), and substantially lowered the price, from around $6 per thousand cubic feet (MCF) to less than $3. What effect do you think this price decline has on the economics of wind development?

# FOCUS: RENEWABLE ENERGY IN HAWAII

Alone among the fifty states, Hawaii cannot call upon its neighbors for electricity. As a result, the state is actively supporting the development of renewable energy resources. Imported oil as of 2012 supplied more than three-quarters of Hawaii's energy needs.[4] No other U.S. state is so critically dependent on imported oil. Environmental and energy security concerns, however, make it essential to sharply reduce oil use in electricity generation and more effectively utilize local sources of renewable energy: solar, wind, water, wave, ocean thermal, biomass, and geothermal.

---

[4] www.hawaii.gov.

While as of 2004 the state obtained less than 0.1 percent of its electricity from wind and 2 percent from geothermal, by 2011 the state produced 12% of its electricity from renewables, and had a firm goal of moving to 40% from renewables (wind, solar, and geothermal) by 2030.[5]

Here is a state assessment of the attributes of renewable energy:[6]

1. Renewable energy projects create more jobs than a comparably sized fossil fuel plant.
2. A greater use of Hawaii's abundant renewable energy resources would help to insulate the state against fossil fuel price escalation and supply disruptions.
3. Money spent on indigenous energy would largely remain in the state.

There are also substantial environmental advantages to renewable energy. Reducing oil use would reduce oil spill risks. This is important to the state's tourist industry, as well as to its extensive marine nature reserves.

But there are also limitations. These include the high cost of land, as well as technical limitations in the state's electricity distribution system. Here is the state's assessment of the overall potential.

Renewable energy projects can theoretically provide all *new* generation required to satisfy projected electricity demand increases in the state through 2014. On Oahu, where most people live, using optimistic assumptions, renewable energy projects could meet all of the electricity required to meet projected demand.

The assumptions usually refer to reasonable interest costs, land lease costs, zoning requirements, construction costs, etc. Although the state considers that wind projects might displace agriculture and thus might not be viable, many wind sites are presently operated on sites compatible with agricultural activity. However, in Hawaii, the two main crops, sugar and pineapples, require harvesting techniques that are not as compatible with wind energy as other crops. With pineapple cultivation, for example, the location of wind turbine towers could interfere with the equipment used to harvest pineapples. The burning of cane fields could also damage wind turbines, but this also represents a biomass residue that could be burned for fuel or converted to ethanol.[7]

Some of the best sites for wind development are offshore. This could make it easier to obtain lease sites, assuming the tourist industry were to concur.

One factor that may ultimately decide the issue could be the future price of oil. Since the state report we cite here was issued, oil prices have nearly doubled, to around $100 a barrel, on world markets.

**Question 10-18:**    What role should renewables play in a sustainable society?

---

[5] www.hei.com.
[6] http://hawaii.gov/dbedt/main/whats_new.
[7] Ibid.

**Question 10-19:** Summarize the important points of this Issue.

## FOR FURTHER THOUGHT

**Question 10-20:** Another proven sustainable source of electricity is geothermal energy. Research and assess the state of geothermal energy development in the United States. Assess the future potential for expanding this resource.

**Question 10-21:** Solar photovoltaic technology employs another renewable energy source: the sun. Research the growth of the photovoltaic industry by searching the web, using the term "solar photovoltaics." Discuss the potential for photovoltaics development over the next several decades.

**Question 10-22:** When you drive or fly, you can offset your greenhouse gas emissions by purchasing greentags, also known as renewable energy certificates. The money you pay goes to develop alternative energy projects (typically wind or solar). The Green Power Network publishes a table listing greentag providers and comparing price premiums for residential greentags.[8] Some environmentalists consider greentags counterproductive, feel-good environmentalism. Research the issue and discuss why this is so and whether you agree.

**Question 10-23:** Hybrid cars are becoming popular in the United States, but the present generation permits few of the cars to be "plugged in" at night or when parked to recharge batteries. Experts claim this improvement could double city mileage for hybrids. Research this issue, and determine why manufacturers like Toyota are reluctant to sell cars that can be plugged in.

**Question 10-24:** Evaluate the extent to which wind potential is being used to generate electricity in your state or province. If your state or province does not have any wind developed, find out why. What sources produce your state's or province's electricity? Coal? Nuclear? Gas?

**Question 10-25:** Much future renewable energy development depends in part on the future world price of oil and natural gas. You have seen what has happened to natural gas prices in the U.S. due to "fracking" production. How predictable is the future oil price? Go to www.eia.doe.gov and determine how the government projects the world price of oil. Some have argued that oil-producing countries could sabotage the development of renewable energy by periodically dropping the price low enough to render renewables "uncompetitive" with oil. Is this a reasonable possibility? Why or why not?

---

[8] www.eere.energy.gov/greenpower/markets/certificates.shtml?page=1.

# Issue 11

## GLOBAL WATER SUPPLIES:
## *ARE THEY SUSTAINABLE?*

### KEY QUESTIONS

- How much water do people need?
- How abundant are world supplies of clean water?
- How much water is used in agriculture?
- Who provides the world's water?
- What effects will population growth and development have on water supplies?
- How could disagreements over water lead to conflicts among nations?
- What effects will climate change have on water supplies?

### INTRODUCTION

In 1977, California was in the midst of a powerful drought. Residents were forbidden to water their lawns or wash their cars. Children learned a rule governing toilet flushing, "If it's yellow let it mellow." One of the greatest hardships the state's middle-class residents faced was a prohibition against washing driveways—a popular weekend activity. Springing to the rescue, entrepreneurs marketed a heretofore rarely used Japanese invention—the leaf blower. By 1980, the leaf blower had found its way into California's (and America's) heart and all because of a water shortage. It is now one of the top sources of air pollution in the state.

Population growth and looming climate change threaten to alter weather patterns and could materially reduce California's access to fresh water. Is this a water crisis?

Half a globe away, Bangladesh faces another kind of crisis: arsenic contamination of the shallow aquifers on which many Bangladeshis rely. Health officials have shown that cholera, an often deadly bacterial infection, can be controlled by a simple and cheap cloth-filtration system for drinking water, but arsenic cannot be so easily removed. Is this a water crisis?

### GLOBAL WATER USE

Throughout the 20th century, while human population tripled, water use increased by a factor of six, according to the World Water Council.[1] Population is projected to increase by 30–40% by 2050. How will this affect global supplies, and global water demand?

Global water use is shown in Figure 11-1.

**Question 11-1:** Which economic sector—agriculture, industry, or municipalities—uses the most water?

---

[1] World Water Council, www.worldwatercouncil.org/index.php?id=25.

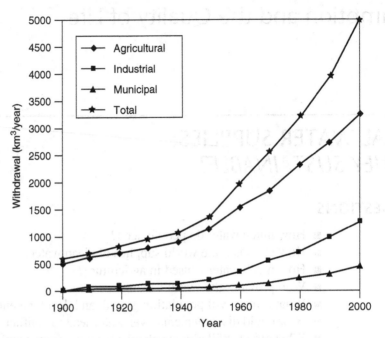

FIGURE 11-1 Global water use by sector. (Worldwatch Institute, *Water in Crisis*, www.worldwatch.org. Courtesy of World Watch Magazine.)

Nearly one in six of the planet's 7.0 billion people presently lack access to safe water, and at least 2.4 billion do not have access to adequate sanitation. Nearly 1.5 million children die needlessly each year because of preventable water-borne diseases. Water demand is projected to grow by as much as 40 percent over the next twenty years. If this issue is to be addressed, hundreds of billions of dollars must be invested globally in water infrastructure over the next five decades.

We begin investigating global water issues with a review of the hydrologic cycle.

# THE HYDROLOGIC CYCLE

Water at the Earth's surface moves through three *states*—liquid, solid, and vapor—and is carried from place to place at the surface of the Earth. The movement of water at the Earth's surface is called the *water,* or *hydrologic, cycle.* Figure 11-2 shows the water cycle. The labeled components of the cycle are *evaporation, precipitation, runoff,* and *infiltration.* Water that infiltrates into pores in soil, sediment, or rock is called *groundwater,* where it may be stored for millennia. Water may similarly be stored for long periods as ice. Table 11-1 shows where the water is found.

**Question 11-2:** Where is most fresh water at the Earth's surface?

**Question 11-3:** How could climate change affect the distribution of water in Table 11-1?

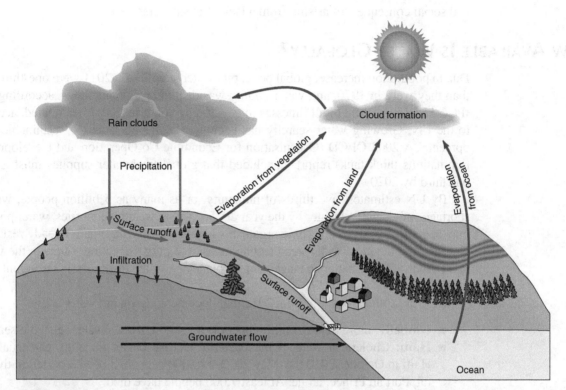

**FIGURE 11-2** The hydrologic cycle. (Keller, E. A. 2006. Natural Hazards. Fig. 1.13, page 15 Pearson Prentice Hall, Upper Saddle River, NJ. Courtesy of E. A. Keller/Pearson Education.)

**TABLE 11-1** ■ Where Is the Water?

| | As a Percent of All Water | As a Percent of Global Fresh Water |
|---|---|---|
| Earth's Oceans | 97.50% | — |
| Glaciers & Permanent Snow | 1.74% | 68.70% |
| Fresh Groundwater | 0.76% | 30.10% |
| Salt Groundwater | 0.99% | |
| Ground Ice & Permafrost | 0.02% | 0.86% |
| Fresh Water Lakes | 0.007% | 0.25% |
| Salt Water Lakes | 0.006% | |
| Soil Moisture | 0.001% | 0.05% |
| Atmosphere | 0.001% | 0.04% |
| Swamps | 0.0008% | 0.03% |
| Rivers | 0.0002% | 0.006% |
| Plants | 0.0001% | 0.003% |

## HOW MUCH IS WATER WORTH?

One observer said that while we might think that diamonds are more valuable than water, it really depends on how thirsty we are. Most things that have economic value have substitutes to which users can turn if the price becomes too high: natural gas for oil, mass transit for autos, spinach for broccoli, etc. But there is no substitute for water. Moreover, the demand for water is growing with human population and, locally, with affluence. Until recently, most people in wealthy countries thought little about the ecological, economic, and social consequences arising from a lack of clean water.

# HOW AVAILABLE IS WATER GLOBALLY?

Due to population increase, global per capita water supplies by 2011 were one-third lower than they were in 1970, and water *quality* was declining in many areas, according to the IPCC.[2] Eighty percent of all illnesses in developing countries are water-related, according to the UN. Growing water scarcity has become a major obstacle to "sustainable development." A 2002 OECD (Organisation for Economic Co-Operation and Development: a rich-nations think-tank) report[3] concluded that global freshwater supplies must grow by one third by 2020.

By UN estimates, two-thirds of humanity, or as many as 5 billion people, will face shortages of clean freshwater by the year 2025. Even in wealthy countries, water problems may become serious. Some toxic organisms like *Cryptosporidium* are already resistant to chlorination, the most widespread technique used to purify drinking water. In the view of many hydrologists, better management of planetary water resources is an urgent global need.

Here are some snapshots from 2012 that document a growing water crisis:[4]

■ **Globally**: more than 3.5 million people a year die from a water-related disease.
■ **Haiti:** Cholera, a water-borne bacterial disease began ravaging earthquake-torn Haiti in October 2010. By May 2012, 140,000 cases had been reported in the capital, Port au Prince, alone. At least 7,000 people have died.
■ **China:** Beijing's 20 million residents have faced increasing water shortages since 1999. By 2011, per capita water supplies had fallen to 100 cubic meters a year, one-twentieth of the national average. Groundwater was down by 37% and surface supplies by 59%.[5]

## Are Water Conflicts Ahead?

On the global stage, at least 260 rivers—the Danube, the Volga, the Ganges, the Brahmaputra, and the Mekong, to name a few—flow through two or more countries. Likewise, neither lake basins nor groundwater aquifers recognize national boundaries. Three-fifths of the world's population lives in the watersheds of international freshwater systems. No global doctrine governs the allocation and use of these international water bodies. Even in the United States, water doctrines and laws vary from state to state. We revisit this subject in "For Further Thought" at the end of the Issue.

# WHO COLLECTS AND DISTRIBUTES THE PLANET'S WATER?

Sizeable investments are required to extract, purify, and distribute water. Should economic and population forecasts materialize, the UN estimates that at least US $180 billion will be required globally to expand supplies over the next two decades. This estimate does *not* include what will be required to rehabilitate or modernize existing systems, nor does it include operation and maintenance costs.

---

[2] Intergovernmental Panel on Climate Change. See for example www.ipcc.ch/pdf/assessment-report/ar4/wg1/ar4-wg1-spm.pdf.

[3] OECD, http://www.oecd-ilibrary.org/economics/oecd-annual-report-2002_annrep-2002-en.

[4] www.water.org/resources/headlines.htm.

[5] *China Times*, 3/15/2012.

There are a number of points you need to know to better understand water issues.[6]

1. First and foremost, *waste* is endemic in the system. Leaking pipes, evaporation from reservoirs, inefficient irrigation and household practices, as well as pollution, all waste water on a vast scale. For example, a substantial part of New York City's water supply disappears unmetered, presumably through leaking underground pipes. And unlined, uncovered irrigation canals can lose more than a quarter of their water through evaporation and seepage.

2. The benefits of enough fresh water extend far beyond anything measurable by economics. For example, were lack of water or polluted supplies to generate large numbers of refugees, the costs would be borne by other segments of society or other nations.

3. Water distribution systems are natural *monopolies*, since it is impractical to have numerous, competing irrigation systems or public water supply systems.

**Question 11-4:** Does our statement in #3 above violate American "free market" principles? Is it important to force competition on the global water system? Discuss and cite evidence for your position.

4. The pricing of water is usually inefficient. Governments typically build and operate water distribution systems, and the prices charged are often not market-based. Charging a price for water that includes all costs is politically difficult. Politicians in wealthy democracies are often fearful of offending large-scale water users, like irrigation districts in California. And riots may occur in poor countries if water prices are raised.

5. Finally, although water circulates globally, its use by humans is local. Thus, water problems are best dealt with on the regional to local scale. However, national and international agreements are an essential framework for local and regional solutions. "All of us live downstream" means that, since water circulates globally, we are all vulnerable to the effects of water pollution.

## IMPACTS OF CONTAMINATED WATER

According to the UN, water scarcity is one of the major factors driving mass migration, creating increasing numbers of human refugees.[7] Water pollution impacts human health in three main ways.

■ *First,* humans need access to a minimum amount of clean water—at least 50 cubic meters a year—both to ensure physical survival and to meet minimal hygiene demands.

---

[6] OECD Factbook, 2010, www.oecd.org/.
[7] United Nations, http://www.un.org/waterforlifedecade/scarcity.shtml.

**Question 11-5:** How does this number compare to the water available to the average resident in China's capital, cited above?

**Question 11-6:** How much water is used in the average shower, or to flush a toilet? (For comparison, the average person in the United States uses about 180 gallons per day—about 680 L).

- *Second,* drinking or bathing in water containing animal or human waste facilitates transmission and proliferation of disease-bearing organisms. The most common waterborne infectious and parasitic diseases include hepatitis A, cholera, typhoid, roundworm, guinea worm, leptospirosis, and schistosomiasis. In developing countries, according to the World Bank,[8] diarrheal diseases alone cause an estimated 3 million deaths and 900 million episodes of illness annually, mainly affecting children. We will revisit this topic in "For Further Thought" at the end of this Issue.
- *Third,* surface water and groundwater can dissolve and/or transport inorganic and organic chemicals, heavy metals, and other toxins. These can cause illness, cancer, birth defects and other mutations and can impair immune system function as a result of direct (drinking contaminated water) or indirect (eating plants or animals harvested from contaminated water) exposure.

## WATER USE IN THE UNITED STATES

Figure 11-3 shows water use in the United States through 2005, the last year for which data were available when we went to press. The United States uses about 410 billion gallons each day. This figure includes both *consumptive* use (the water is not put back where it came from) and nonconsumptive use (the water is put back after use).

**Question 11-7:** Describe changes in U.S. per capita water use over the period 1975–2005.

---

[8] World Bank, World Development Report, www.worldbank.org.

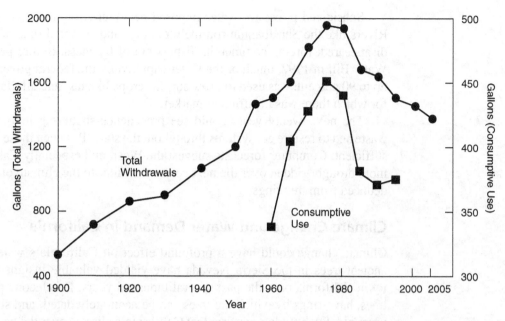

**FIGURE 11-3** U.S. water use to 2005, per capita, per day. *Consumptive use* refers to water that is removed and not returned to a system, e.g. agricultural and manufacturing uses.

## Focus: Water Use in California

In the United States, water plays a critical role in nearly every state west of the Mississippi River. California is a great example. Most of the state has a Mediterranean or desert climate, and rain falls mainly between October and April.

California is home to 38 million people and adding upwards of 300,000 new residents a year. Up to three-fourths of the water used in California, or about 34 million acre-feet, is by agriculture, even though agriculture provides only about 4 percent of the state's GDP and 1 percent of its jobs. Water has been heavily subsidized to agribusinesses for decades. In parts of the Central Valley, 1 acre-foot (325,000 gal. or about 1228 m$^3$) could be purchased in 2005 for about $7.50 (a rate of about 770 L for a penny). At the same time, farmers in San Diego County were paying $400 per acre-foot, and residents of Santa Barbara faced a charge of $1,900 per acre-foot.

**Question 11-8:** At $1900 per acre-foot (325,000 gallons) how many gallons could be bought for a dollar?

Until recently, in many California cities, water was viewed as "too cheap to meter." Even during droughts, users paid a flat fee meaning conservation couldn't be legislated.

Subsidized prices to agribusiness meant massive waste. Most irrigation ditches were made of unlined dirt, which meant that up to a quarter of the water shipped from California's rivers to farm fields trickled into the soil before reaching its destination. Up to 25 percent more evaporated from ditches in 100-degree heat or from reservoirs behind vast dams.

Subsidized prices for water imposed heavy environmental costs on wildlife as well. Rivers like the San Joaquin routinely ran dry, and seasonal runs of salmon and steelhead disappeared, forcing commercial fishers out of business. Before passage of the Omnibus Water Bill in 1992, much of the water supplied by the federal government, at subsidies of up to 90 percent, was used to grow surplus crops like hay and alfalfa; in other words, crops for which there was insufficient market.

The next twenty years could see precedence-shattering innovation to reduce water waste and to restore ecosystems throughout the state. But even those measures might not be sufficient. Computer forecasts suggest the West, and especially California, could become more drought-ridden over the next century, similar to the climate of the prehistoric past as deduced from tree rings.

## Climate Change and Water Demand in California

Climate change could have a profound effect on California's water supplies. Studies of ancient trees in the Sierra Nevada have yielded valuable insight into climatic variability in California over the past several thousand years. Bristlecone pines, the oldest living trees, have rings by which the trees can be accurately dated, and some of them are 4,000 years old. Such studies confirm that California's climate over the past millennium has been extremely variable, with severe drought conditions persisting for decades to centuries.

We can now incorporate this historical record into computer forecasts of future climates based on changing levels of greenhouse gases. California's climate is "forecast" (see Issues 5 and 6) to become more extreme than in the recent past. Should such forecasts, or something even more severe, become the norm, these climatic extremes will critically stress the state's water storage and delivery system and may force agonizing choices between protecting aquatic ecosystems and providing the massive amounts of water consumed by industrial agriculture, commerce, and municipalities.

**Question 11-9:**    Study Table 11-2. How does California compare to the other cities in the U.S.? To Australia? How does urban use compare to the minimal human water requirement of 50,000 liters per year (1 gal. = 3.8l)?

**Question 11-10:**    California's population could top 50 million by mid-century. Discuss the implications for sustainable water use under this scenario.

TABLE 11-2 ■ Water Use in the Western US and Australian Cities (Units are gallons per capita per day [GCPD])

| Location | Residential Use** (gpcd) | Urban Use** (gpcd) |
|---|---|---|
| Portland, OR | 60 | 116 |
| Albuquerque, NM | 74 | 154 |
| Tucson, AZ | 97 | 144 |
| Denver, CO | 104 | 160 |
| **California** | **111** | **162** |
| San Francisco | 54–56* | 95–102* |
| Los Angeles | 77*–107 | 139*–154 |
| San Diego | 79*–113 | 136*–157 |
| Oakland/East Bay | 87*–100 | 138*–146 |
| San Jose | 91*–97 | 156–160* |
| Sacramento | 93–128* | 142–247* |
| **Australia** | **63** | **100** |
| Melbourne | 53 | 87 |
| Sydney | 56 | 90 |
| Canberra | 61 | 95 |
| Brisbane | 74 | 122 |
| Perth | 76 | 110 |

Notes
* From Urban Water Management Plan.
** Does not include unaccounted for water (e.g., system leaks)

# WATER AND DEVELOPMENT: THE PEOPLE'S REPUBLIC OF CHINA

In early stages of economic development, little investment is placed in maintaining environmental quality. "Modern" diseases such as cancer and diabetes and "epidemics" such as obesity and tobacco use generally increase alongside the decline of traditional maladies, like waterborne infectious diseases. This is the "Risk Transition" concept.

According to the World Resources Institute,[9] the Risk Transition interval is illustrated by the recent history of China. Declining but still substantial traditional illnesses and increasing modern ailments pose a major challenge to China's health-care system. The relationship between water and health will strongly influence China's present and future development.

Approximately 700 million Chinese—over half the population—must use water that contains levels of animal and human waste that fail to meet minimum drinking water quality standards. Three reasons account for China's degraded water:

1. Rapid and unregulated expansion of industry
2. Failure to invest in infrastructure to meet growing urban water needs
3. Reliance on sewage effluent to irrigate crops

Industries scattered throughout rural China are gradually replacing obsolete state-owned concerns.

These industries are unregulated and have few if any pollution-control facilities. As a result, such contaminants as excess organic matter, acids, alkalis, the nutrients nitrogen and phosphate, organic and inorganic chemicals, and heavy metals such as lead, cadmium,

---

[9] World Resources Institute, www.wri.org.

mercury, and chromium are among the major water pollutants detected in rural water bodies. The impacts of these toxins fall heaviest on the very young and very old.

Urban areas likewise experience heavy pollution loads. Each year, more than 30 billion tons of urban sewage is discharged into rivers, lakes, or seas with virtually none receiving even rudimentary treatment. Chinese researchers report that untreated sewage usually contains dangerous levels of pathogenic microorganisms. Rivers in cities are also polluted by inorganic and organic chemicals, many of them known to be carcinogenic, mutagenic, or both. Some researchers found associations between gastric diseases and elevated cancer rates in areas irrigated with sewage as well as industrial wastewater. When municipal water is chlorinated (to kill microbial pathogens), halomethanes such as chloroform and trihalomethane are formed by the reaction between chlorine and organic matter. These substances in high enough concentrations may pose a serious cancer risk (up to one in one hundred).

Agriculture is a major source of water pollution in China. Since the 1980s, many farmers have adopted so-called "green revolution" technologies common to western industrial agriculture: intensive fertilizer and pesticide application together with hybrid seeds. As a result, the use of pesticides and fertilizers has been increasing in China. Fertilizer use increased by at least one-third during the 1990s, but inefficient application meant that up to 70 percent of the fertilizer was wasted, leading to high levels of toxic nitrate in groundwater, which can kill infants.

Yet while water pollution is a persistent threat, more than half of Chinese cities experience severe water *shortages,* which we illustrated above with Beijing. These could also be called demand *longages.*

**Question 11-11:** Could water shortages and water pollution be a cause for the increase in Chinese nationals trying to gain entry into the United States and other countries? Why or why not?

**Question 11-12:** Are cheap imports of consumer goods from China being subsidized by insufficient investment in environmental protection in that country? What information would you like to have in order to answer the question more fully?

# WATER AND GEOPOLITICS: ISRAELI/PALESTINIAN WATER CONFLICT

Since the advent of Jewish repatriation to Palestine around the turn of the twentieth century, conflicts over the region's scarce water have been growing. They continue to pose one of the thorniest problems stalling a comprehensive Middle East peace. At the center of this conflict is the disposition of groundwater underlying the West Bank region, called Judea and Samaria by the Israelis, access to which is claimed by both Israelis and Palestinians.

**TABLE 11-3 ■** The Safe Yields of the Three Main Aquifers in the Middle East

| Aquifer system | Number of aquifers | Annual safe yield (cubic meters × $10^6$/y) |
|---|---|---|
| Western Basin | 2 | 350 |
| Northeastern Basin | 2 | 140 |
| Eastern Basin | 6 | 125 |

Further complicating the issue is the rapid population growth in the region on both sides, on the Israeli side coming substantially from immigration.

Groundwater supplies in the region are shown in Table 11-3. Groundwater is the most important source of fresh water in the region and comes mainly from aquifers located and recharged in the West Bank. Total precipitation is estimated at 2,600 millimeters, of which only about 600 millimeters per year is infiltration.

Untapped groundwater supplies as of 1998 were estimated at 78 million cubic meters (mcm) per year.

## Water Supplies in Gaza

Water supplies in the Gaza Strip are even more restricted than supplies elsewhere. Other than home cisterns to catch and store scarce rainfall, groundwater is the only source of fresh water. The present safe yield of the aquifers is estimated at 65 million cubic meters per year (mcm/y), but at least 100 mcm/y is being withdrawn. Thus, these systems are being "mined" at the rate of at least 35 mcm/y. As they are in contact with seawater, the aquifers are being poisoned by saltwater intrusion and will eventually be unusable should present trends continue. Since the population of Gaza is increasing, these trends will likely worsen. As of 1995, over 60 percent of groundwater withdrawals went to irrigate crops, so should irrigation efficiencies increase, more supply would be available to protect the system from saltwater incursion.

## West Bank Groundwater

The Judean/West Bank aquifer system underlies the Palestinian West Bank, the Jordan River/Dead Sea Rift, and the narrow Israeli coastal plain. Water was one of the major items of disagreement between Israelis and Palestinians during the negotiations leading to the interim "Oslo B" agreement, signed in Washington, D.C., in 1995. While Israel recognized Palestinian water rights in the West Bank, the interim agreement left details unresolved. The CIA estimated the Palestinian population in West Bank and Gaza at 4.3 million in 2012.

Based on present water allocations, the average Palestinian has access to less than 25 cubic meters of water per year.

**Question 11-13:** How does this figure compare to the annual water needed to preserve human health cited earlier?

Were a goal to be set to allocate to each Palestinian the minimum of 50 cubic meters per year recognized to meet basic domestic needs, more than 70 mcm/y of supply would have to be found in a region that is already near capacity in terms of water use.

Approximately 25 percent of Israel's water supply comes from West Bank aquifers. The present allocation of groundwater is described as "the abilities of the strong to impose their wills on the weak."[10] Furthermore, Israeli surface-water use has polluted the Jordan River downstream from Israeli diversions, rendering the water unsuitable for West Bank farmers.

**Question 11-14:**   Do the Israelis have any responsibility to provide adequate water supplies to Palestinians in Gaza and the West Bank? Explain reasons for your answer. Does your answer exacerbate tensions in the Middle East or relieve them? Why?

Future water needs cannot be met by increasing supply from presently existing sources. Thus, conservation, new supplies, or a combination of both will be needed, along with measures to address population growth, if the issue of Middle East water allocations is to be fairly settled. And if the issue is not fairly settled, it is unlikely that the region will experience a long-term decline in hostilities.

**Question 11-15:**   Summarize the main points of this Issue.

**Question 11-16:**   Discuss whether the water demand in the Middle East is sustainable. If it is not, what could be done to make it sustainable?

## FOR FURTHER THOUGHT

**Question 11-17:**   Do an Internet search for anecdotal illustrations of the nature of conflicts over water access and quality. How do authorities propose to address these issues?

**Question 11-18:**   Do a similar Internet search for examples of water conflicts—both international and intranational. Summarize your findings in a paragraph or two.

---

[10] Environmental Degradation and the Israeli-Palestinian Conflict, http://www.arij.org/.

**Question 11-19:** Find examples of water waste either in the United States or abroad. You could focus on agriculture, industrial, or domestic use. Summarize in a paragraph or two what you found.

**Question 11-20:** Do another Internet search using the term "water pricing." Cite examples of irrational pricing or conflicts over water prices.

**Question 11-21:** Find out how your state or province adjudicates water conflicts or regulates "ownership" of water. Summarize in a paragraph what you found.

**Question 11-22:** Pick one of the waterborne diseases mentioned in this Issue and find out more about it. What causes it? How can it be prevented or treated? What is the annual cost of the disease to a country or globally in lives or money? Summarize in a paragraph or two what you found.

**Question 11-23:** The film Erin Brockovich, based on a book, concerns the effects of chromium in a community's water supply. Watch the film or read the book and record your responses.

**Question 11-24:** Watch the documentaries *Flow* (2008) and/or *Tapped* (2010), which focus on privatization of water supplies and whether access to clean water is a human right or a commodity. Use evidence from the videos as the basis to discuss the costs and benefits of privatization of water supplies, and whether you support it.

# MOTOR VEHICLES AND THE ENVIRONMENT I

## KEY QUESTIONS

- What are the major motor vehicle trends in the United States?
- What are the environmental impacts of motor vehicles?
- What are CAFE standards? Are they working?
- What are hybrids, and what is their long-term significance?
- Are personal motor vehicles compatible with a sustainable society?

## BACKGROUND

The model year 2000[1] saw the introduction of the Honda Insight (since discontinued, then reinstated in 2010) and the Toyota Prius (Figure 12-1), the first "hybrid" motor vehicles. These vehicles use gasoline engines, electric motors, and advanced technology to achieve mileages of close to 50 miles per gallon, are also extremely low-polluting vehicles. The Prius was a sales phenomenon. See Table 12-1.

**Question 12-1:** In what year did hybrid sales peak?

**FIGURE 12-1** A 2012 Toyota Prius. More than one million Priuses have been sold in the U.S. since 2000. (Courtesy of Alan Look/Icon SMI/Corbis Images.)

---

[1] A model year begins on October 1 and ends on September 30.

**TABLE 12-1** ■ Sales of Hybrid Vehicles in the U.S.
(Ref: www.hybridcars.com and Statistical Abstract of the U.S.,
www.census.gov)

| Model year | Sales |
|---|---|
| 2000 | 9,350 |
| 2001 | 20,287 |
| 2002 | 35,000 |
| 2003 | 47,525 |
| 2004 | 88,000 |
| 2005 | 205,749 |
| 2006 | 252,636 |
| 2007 | 352,000 |
| 2008 | 312,000 |
| 2009 | 290,000 |
| 2010 | 274,000 |
| 2011 | 268,000 |

**Question 12-2:** Why do you think hybrid sales declined beginning in 2008? [Hint: How did the overall auto market behave?]

With unprecedented nominal (unadjusted for inflation) gas prices exceeding $4/gallon by 2012, the automobile culture, in the view of some observers, is beginning to lose some of its allure, especially among the young. Are we *still* what we drive?

# THE ENVIRONMENTAL (AND SOCIAL) COST AF MOTOR VEHICLES

The environmental and social impact of our reliance on internal-combustion-engine-powered motor vehicles is profound. In the United States, these impacts include the following:[2]

- Health-care costs in the billions associated with crashes[3]
- Annual interest cost near $20 billion to finance the purchase of passenger cars
- Hundreds of millions of vertebrates, including thousands of deer, killed each year. Deer crashes are a major source of auto insurance payouts.
- Air and water pollution (see below), including a major contribution to what the Natural Resources Defense Council calls "poison runoff." Poison runoff includes cadmium, zinc, and other heavy metals from tire wear and hydrocarbons from fuel and grease.[4]
- Accelerated urban sprawl as roads are paved
- Fragmentation of habitat by road building, introduction of alien species, etc.

---

[2] For details, consult websites of these organizations: American Automobile Association, U.S. EPA, Natural Resources Defense Council, Wildlands CPR, Statistical Abstracts.

[3] Insurance Institute for Highway Safety data for 1999 (www.hwysafety.org/safety_facts/safety.htm).

[4] Poison runoff also originates from nonpoint sources (like golf courses) as opposed to point sources such as sewer pipes or factory smoke stacks.

- Dependence on foreign oil from undesirable sources
- Road rage incidents

Here are some additional auto-facts:[5]

- In 2009, there were 10.8 million motor vehicle accidents, resulting in 35,900 deaths.
- In 2009, the average vehicle was driven 11,600 miles.
- In 2009, the U.S. consumed 168.9 billion gallons of fuel for motor vehicles.
- In 2009, urban road congestion cost the average driver $591, each wasting 20 gallons of fuel.
- In 2009, according to the Tax Foundation, the average household spent $216 a month on gasoline and motor oil.

**Question 12-3:** Do you own a motor vehicle? How similar is the average monthly expenditure on gas and oil to your experience?

## WHAT WE DRIVE

Table 12-2 shows sales of motor vehicle by class from 1975 to 2010.

**TABLE 12-2** ■ New Motor Vehicle Sales and Leases, 1975–2010, Millions. (Source: Statistical Abstracts of the US)

| Year | Car | Light Truck |
|------|--------|-------------|
| 1975 | 8.624 | 2.468 |
| 1980 | 8.979 | 2.623 |
| 1985 | 11.043 | 5.193 |
| 1990 | 9.301 | 5.256 |
| 1995 | 8.635 | 7.869 |
| 2000 | 8.847 | 11.529 |
| 2005 | 8.614 | 12.976 |
| 2006 | 7.821 | 8.683 |
| 2007 | 7.618 | 8.842 |
| 2008 | 6.814 | 6.680 |
| 2009 | 5.456 | 5.145 |
| 2010 | 5.729 | 6.044 |

---

[5] Statistical Abstracts of the U.S., www.census.gov.

**Question 12-4:** On the axes below, separately plot sales of each of the above categories for the years 1975, 1985, 1995, 2005, and 2010.

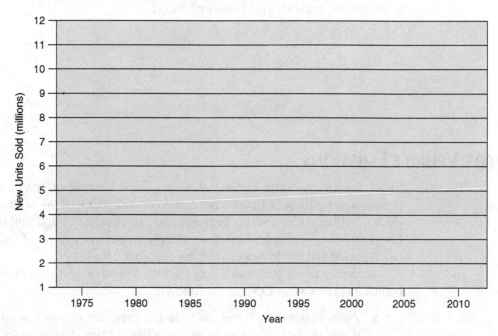

**Question 12-5:** Calculate the percentage of light trucks (including SUVs) out of total vehicle sales and fill in the table below.

| Year | % Light Trucks |
|------|----------------|
| 1975 | |
| 1980 | |
| 1985 | |
| 1990 | |
| 1995 | |
| 2000 | |
| 2005 | |
| 2010 | |

**Question 12-6:** Describe trends in sales of light trucks and passenger cars from the data in Table 12-2 and your graph.

One reason that families buy large vehicles like SUVs is laws requiring large car seats for children and infants. Most vehicles simply cannot accommodate more than two of these behemoths.

**Question 12-7:** List other reasons why people buy SUVs and large trucks. Identify the assumptions you bring to the issue of the "suitability" of these vehicles, and then critically evaluate the reasons you have just listed.

## MOTOR VEHICLE EMISSIONS

Motor vehicles generate significant air pollution because of combustion or evaporation of fuel. Among these pollutants are carbon dioxide ($CO_2$), hydrocarbons, nitrogen oxides (NOx), sulfur oxides (SOx), benzene, soot (particulates), and carbon monoxide (CO). Emissions of new vehicles have been sharply reduced as a result of federal and state legislation passed and implemented over the past four decades.

According to the National Safety Council and other agencies, impacts of each class of pollutant (exclusive of $CO_2$) are as follows:

■ *Hydrocarbons* react with NOx in the presence of sunlight to form ground-level ozone. Higher temperatures enhance the reaction. Exposure can lead to permanent lung damage, among other things.
■ NOx contributes to the formation of ozone and acid precipitation.
■ SOx contributes to acid precipitation and is toxic at high concentrations. It results from oxidation of sulfur in motor vehicle fuels. In 2006, refiners began to reduce sulfur by as much as 97 percent in diesel fuel. By 2010, diesel for on-road trucks contained only 15 ppm sulfur, according to the U.S. EPA.
■ Exposure to benzene, a component of gasoline, can cause cancer and other diseases.
■ "Soot" results mainly from diesel vehicles. Particulates are the major source of urban air pollution in many areas.
■ CO is a colorless, odorless, lethal gas. CO reduces the flow of oxygen in the bloodstream and can impair mental functions and motor response. In urban areas, motor vehicles generate up to 90 percent of atmospheric CO.

Petroleum combustion, mainly for transportation, accounted for nearly 43% of the United States' annual $CO_2$ emissions in 2009 according to the U.S. EIA. $CO_2$ is the major global agent of human-induced climate change. The U.S., in turn, is responsible for 20% of global $CO_2$ emissions.

**Question 12-8:** What are environmental and health impacts if emissions are not reduced? Who pays these costs?

**Question 12-9:** Should people who choose to drive low-mileage vehicles with high emissions pay these costs? How? Be sure to assess your reasoning, using critical thinking principles.

## FUEL ECONOMY AND CAFE STANDARDS

The Energy Policy and Conservation Act of 1975 established fuel economy standards for passenger cars. CAFE (Corporate Average Fuel Economy) standards were implemented for 1980 model year cars at 18.0 miles per gallon (mpg). The CAFE standards for cars and light trucks in 2000 were 27.5 and 20.7 mpg, respectively. New standards for the years 2012–2016 will raise average fuel economy for cars to 35.5 mpg and set tailpipe emissions of $CO_2$ to an average of 250 g/m. For trucks, the standard is 28.8 mpg. In 2011 standards were proposed for 2017–2025, which would gradually raise fuel economy to 54.5 mpg for the 2025 model year. The U.S. EPA estimates that the average cost for compliance will add about $800 to the average cost of a vehicle, but the owner will save about $4000 over the vehicle's life.[6]

**Question 12-10:** The auto industry often responded to CAFE standards by reducing the weight of some vehicles. How could improving fuel economy this way lead to decreased safety?

**Question 12-11:** Is the assumption that increasing fuel economy must be achieved only through vehicle "downsizing" valid? Research the issue. What are other ways in which increased fuel economy can be achieved?

*Car and Driver* magazine conducted an in-depth analysis of the new regulations in 2010 and concluded, "achieving these goals will require various engine and transmission technologies, as well as improved aerodynamics, tires with lower rolling resistance, and materials that reduce weight. After wading through some 1500 pages of documents, we can say that this overhaul of CAFE was carefully considered, involved extensive input from automakers, and—with the new size-based standards—takes into account customer choice in a way that the old system never did. And if gas prices once again head toward $4, customer demand for fuel economy will likely outstrip these regulations."[7]

---

[6] U.S. Environmental Protection Administration, www.ePa.gov.
[7] Csere, Csaba, *Car and Driver*, May 2010.

**Question 12-12:**     Some critics assail CAFE standards for a variety of reasons. Research this issue and evaluate their objections.

## FOR FURTHER THOUGHT

**Question 12-13:**     We discard over 250 million tires a year. Cadmium and other heavy metals partly from tire wear pose a threat to estuaries like San Francisco Bay and Chesapeake Bay. Tire dust also causes human respiratory problems.[8] Evaluate the environmental impact of tire use and discards. How can we reduce tire impact while personal motor vehicle use increases?

**Question 12-14:**     Discuss to what degree low fuel prices are in the best interests of the American consumer.

**Question 12-15:**     How do low fuel prices affect the nation's security?

**Question 12-16:**     How do low fuel prices affect the economics of substitute fuels like biodiesel or ethanol?

**Question 12-17:**     How do low fuel prices affect decisions by energy companies to seek new supplies?

**Question 12-18:**     In 2004 Ford Motor Co. introduced the Escape hybrid SUV, with mileage ratings of 36 mpg "city," 31 mpg "highway." Research whether sales of the hybrid version have been successful in the United States. Interpret your findings.

**Question 12-19:**     By 2012, several manufacturers were again selling all-electric models, including Nissan and General Motors, and more were planning electrics. Research the advantages and disadvantages of all-electric vehicles.

**Question 12-20:**     Research the impact that battery technology has on the cost and range of electric vehicles.

---

[8] *Rachel's Environment and Health Weekly.* 1995. Tire Dust. #439.

## MOTOR VEHICLES AND THE ENVIRONMENT II:
*GLOBAL TRENDS*

### KEY QUESTIONS

- What are examples of motor vehicle emissions?
- What are the major trends in motor vehicle sales and ownership outside of the United States?
- What impacts will China and other developing countries have on the global motor vehicle fleet and the environment?
- Are there sustainable transportation alternatives to internal-combustion vehicles?

### MOTOR VEHICLE EMISSIONS

We introduced this topic in Issue 12. Burning fossil fuels produces gases and particulate matter. These are collectively called *emissions*. Under the "cleanest" conditions, emissions would be $H_2O$ and $CO_2$. However, since the fuel is burned in air in a car's engine at high temperature and since air is 70 percent nitrogen (N), oxides of nitrogen (NOx) are also produced.

**Question 13-1:** Contrast emissions from an all-electric car to a conventional one.

Sulfur is present in most fuels. Diesel fuel for motor vehicles could contain up to 500 parts per million (ppm) legally as of 2000 but had its sulfur (S) content reduced to 15 ppm in 2006. Gasoline may contain up to 30 ppm sulfur. Thus, oxides of sulfur (SOx) will be produced when the fuel is burned.

Marine diesel engines, which are permitted to burn very high sulfur "bunker fuel" represent the greatest source of diesel emissions for many coastal communities in California, Oregon, Washington, Alaska, and British Columbia. Up to 250 million tonnes of bunker fuel oil is burned each year by ocean-going vessels. In U.S. coastal waters, maximum permitted sulfur in ocean vessels is 1,000 ppm.

**Question 13-2:** Read the following quote from the EPA on the effects of the new low-sulfur diesel rule, and comment on the annual cost savings to the health-care system. One way to begin is to consider the cost for an emergency room visit for a child with asthma, for example.

> Once this action is fully implemented, 2.6 million tons of smog-causing nitrogen oxide emissions will be reduced each year. Soot or particulate matter will be reduced by 110,000 tons a year. An estimated 8,300 premature deaths, 5,500 cases of chronic bronchitis and 17,600 cases of acute bronchitis in children will also be prevented annually. It [will help] avoid more than 360,000 asthma attacks and 386,000 cases of respiratory symptoms in asthmatic children every year. In addition, 1.5 million lost workdays, 7,100 hospital visits and 2,400 emergency room visits for asthma will be prevented.[1]

Carbon monoxide (CO) is produced as well. When properly maintained, catalytic converters convert almost all the CO to $CO_2$. If some of the diesel or gasoline is not completely burned (typical of engines before they are warm), various *volatile organic compounds* (VOCs) are produced. VOCs, CO, NOx, and other gases react with sunlight to produce low-level *ozone* ($O_3$).

So the emissions from motor vehicles may include NOx, SOx, CO, VOCs, ozone, particulates, $H_2O$, and $CO_2$. Let us review the toxic effects of these emissions. NOx and SOx react with moisture in the air and result in *acid precipitation*. Recent research has implicated SOx emissions in the production of methylmercury by bacteria in soils and wetlands (see Issue 17; Methylmercury is the most toxic form of this metal). CO is harmful to organisms and can be lethal at high concentrations. VOCs and fine particles are *carcinogens* (cancer-causing agents). $CO_2$ is a greenhouse gas. Its impact is discussed in Issues 5 and 6. Ozone can harm crops. Low concentrations can cause eye irritation, and high concentrations can damage animal respiratory systems.

## AUTOMOTIVE USE IN CHINA

We will next consider the global economic and environmental impacts of an exponential growth in automobile use worldwide. We focus first on China.

In China as in India, motor vehicle use will probably increase substantially over the next several decades. China's 1.4 billion people are poised to become a global economic superpower in the twenty-first century. It already has an enormous environmental footprint.

First, some historical background: A 1994 article in the *Washington Post* titled "Dreams on Wheels" described a car purchase by members of the tiny but growing Chinese middle class.[2] Mr. Wang was one of only 350,000 Chinese to purchase a car in 1994, but the official newspaper, *China Daily,* enthused over a domestic market consisting of "about 300 million potential car owners." By 2004, government forecasts were slightly lower (at 250 million), but such estimates have historically been conservative.

Approximately 1.2 million personal motor vehicles (PMVs) were sold in China in 2000.[3] Automotive Resources Asia projected sales of 1.976 million per year by 2005.[4] This figure proved too low according to *Ad Age China:* 2005 sales were reported at around

[1] www.epa.gov/OMS/highway-diesel/index.htm.
[2] Mufson, S. Dreams on wheels. *Washington Post,* December 28, 1994.
[3] Smith, C. S. The race begins to build a small car for China. *New York Times,* October 24, 2000.
[4] http://web.archive.org/web/20110128215730/http://auto-resources-asia.com/.

**FIGURE 13-1** A traffic jam, Chinese-style. (D. Abel)

4 million vehicles! But ownership remained low—in 2004 there were 20 PMVs per 1,000 people. There were 770 PMVs per 1,000 people in the United States.

Let us consider the impact of adding millions of cars per year to roads in China and the impact of an eventual fleet of 300 million motor vehicles (Figure 13-1).

By 2010, China was producing 14 million motor vehicles a year, and was the world's largest car and truck market. And in the first quarter of 2012 alone, China produced 4.8 million motor vehicles. The fastest sales growth was in SUVs.[5]

**Question 13-3:** Project sales figures for 2012, then check to see if your projection was correct.

**Question 13-4:** Passenger vehicle sales were 350,000 in 1994 and rose to 14 million in 2010. What was the annual growth rate in sales per year? (Use the formula $r = (1/t) \ln(N/N_0)$, introduced in "Using Math in Environmental Issues," pages 6–8.)

**Question 13-5:** At this rate of increase, how long would it take China to put 300 million vehicles on the road? Use as a starting point 4 million PMVs in 2005. (Use the formula $t = (1/r) \ln(N/N_0)$, introduced on page 7.)

---

[5] www.chinaautoweb.com.

**Question 13-6:**   Assume each car averages 25 mpg[6] and is driven 10,000 miles each year. How much fuel, on average, would each car use per year? Express your answer in gallons and liters. Keep in mind the assumptions under which you are doing this analysis.

**Question 13-7:**   How much fuel (gas or diesel) would be required for 300 million cars annually? What is the daily requirement in barrels per day (1 barrel = 42 U.S. gal)?

As of 2012, global oil production was around 85 million barrels per day.

**Question 13-8:**   The United States in 2010 was the world's largest consumer of gasoline and used around 8.8 million barrels per day (1 barrel = 42 gal) in 2006. Compare this figure with the projected Chinese demand for gasoline.

**Question 13-9:**   At a world price of about $100 a barrel, how much do Americans spend each day on gasoline alone?

**Question 13-10:**   In addition to rising Chinese demand for petroleum, India's 4 percent share of global oil demand (about 3.5 million barrels a day in 2012) was forecast to rise to 10 percent by 2030. What do you think will happen to the price of oil in the face of this increased Indian and Chinese demand? (A good place to find projected future prices of oil is the EIA's Annual Energy Report).[7]

---

[6] Our assumed miles per gallon includes the amount of time each car spends idling in traffic, which in China is profound.
[7] http://www.eia.gov/oiaf/aeo/pdf/0383(2010).pdf.

**Question 13-11:**  Many developing countries, like Indonesia for example, depend heavily on kerosene for cooking, and the price is usually heavily subsidized by government. Discuss the impact of oil price increases on developing countries. List any assumptions you will have to make.

Now, let's consider the environmental impact of 300 million cars. In 2008, China moved to the more stringent Euro III standards (average 0.09 mg/km NOx).[8]

**Question 13-12:**  Assume 0.09 gram NOx per kilometer. How many metric tons of NOx would be generated by 300 million motor vehicles, each driven 16,000 kilometer per year in China? Compare these values to the NOx produced by motor vehicles in the United States, which was 5.2 million tonnes in 2008 (from on-road vehicles alone).[9]

One other factor should be considered: the extent to which diesel vehicles penetrate the Chinese market. In Europe, NOx standards for diesel are more lax than for gasoline engines.

As you can see, emissions from Chinese motor vehicles depend on many factors: emissions regulations, congestion, the diesel-gasoline mix, etc. According to the World Bank, as of 2005 China used more than seven times the energy of the United States per unit of GDP and twelve times as much as Japan.[10] This means, for example, that if China's GDP were the same as that of the United States, China would consume seven times the energy of the United States.

**Question 13-13:**  Comment on the extent to which energy efficiency and conservation investments could help reduce China's air pollution problems.

---

[8] www.dft.gov.uk.
[9] Statistical Abstract of the U.S., http://www.census.gov/compendia/statab/.
[10] www.voanews.com.

## GLOBAL TRANSPORTATION TRENDS

The potential for growth of the motor vehicle fleet worldwide is enormous. By the year 2025, 1 billion vehicles could be operating on the world's roads.[11] While per capita ownership of motor vehicles is low in developing countries, in the case of China and India, it is on the rise. See Table 13-1.

**TABLE 13-1** ■ Car Ownership, Selected Countries per 1,000 inhabitants, 2009 (World Bank, www.worldbank.org.)

| USA | 800 |
|-----------|-----|
| Brazil | 209 |
| Indonesia | 79 |
| China | 47 |
| India | 18 |

While India has low per capita car ownership, it has 1.2 billion people and the world's largest middle class. Indian consumers bought 650,000 motor vehicles in 2003, almost 1 million in 2004, 1.87 million in 2010, and 1.95 million in 2011. (In 2005, Hyundai Motors had projected overall sales at 2 million units by 2010.)[12]

**Question 13-14:** Based on the change in car sales between 2003 and 2011, estimate the doubling time for Indian car sales. (Use t = 70/r; see "Using Math in Environmental Issues," pages 6–8.)

**Question 13-15:** If India were to have the same per capita car ownership ratio as the U.S., how many cars would that involve, based on a population of 1.2 billion? Is that likely? Assess the infrastructure needs of India for a level of car ownership similar to China's today.

There are some hopeful signs (new mass transit in Ecuador, banning of old buses in Delhi and Beijing, the phase-out of leaded gas in Manila, the rapid increase in fuel-efficient car sales) that the increase in motor vehicles in developing countries will be accompanied by environmental regulations, sensitivity to alternative forms of transportation, and some attention to sustainability.

---

[11] World Resources Institute, www.wri.org.
[12] JTE Energy Research Associates, http://www.judeclemente.com/india.

**Question 13-16:** Summarize the main points of this Issue.

**Question 13-17:** Discuss the issue of auto transportation in China and India from the standpoint of sustainable societies.

## FOR FURTHER THOUGHT

**Question 13-18:** Based on 300 million motor vehicles, the Chinese would have to dispose of or recycle several hundred million tires per year. Research the issue posed by tire disposal in the United States and discuss.

**Question 13-19:** Research the impact of fossil fuel emissions on human health in China. Summarize your findings.

**Question 13-20:** Research Chinese plans for hybrid or electric car production, and their attempts to control dangerously high air pollution in Beijing and elsewhere. Compare pollution levels in Chinese cities to those in the U.S. and the EU. What is the source of most of this pollution?

# Issue 14

## WHACKER MADNESS?
### *THE PROLIFERATION OF TURFGRASS*

## KEY QUESTIONS

- What is the economic impact of turfgrass and lawns?
- What are the positive and negative environmental impacts of turfgrass proliferation?
- How significant is pollution caused by turf and lawn maintenance?
- Is water use in turf maintenance a significant environmental issue?
- Are millions of acres of turf consistent with sustainability?

## BACKGROUND

A reader wrote to advice columnist Ann Landers:[1]

> Dear Ann:
>
> We bought a lakefront home in the woods thinking we would get away from the weekend Lawn Rangers, to no avail. Our neighbors have this golf course mentality, which is positively maddening. If it isn't the lawn tractor, it's the weed whacker, the leaf blower, the power mulcher, the lawn vac or a chain saw. . . . Enough already!
>
> We spend a minimal amount of time mowing down the weeds . . . because there are better things to do in life than mow lawns and contribute to noise pollution. They probably refer to us as—
>
> The Schlocky Neighbors in Knowlton, Wisc.

**Question 14-1:** Before continuing, contrast activities involved in "gardening" with those involved in "lawn maintenance."

---

[1] December 19, 1997.

## LAND-USE CHANGES IN NORTH AMERICA

Before the arrival of Europeans, most of the eastern United States was hardwood forest. Changes in land use during the ensuing 400 years have been profound, as documented in sediment cores taken from Chesapeake Bay[2] (see Issue 21). The first major change involved clearing of old-growth forest for agriculture, which was substantially complete by 1920, at least in the watershed of Chesapeake Bay.[3]

Land-use changes continue, most notably the "suburbanization" of much of the United States, which was a hallmark of the twentieth century. Accompanying this profound shift in land use has been the conversion of large areas of the United States to turfgrass.

We owe to the twentieth century the idea of a "smooth, green carpet as a necessary adjunct to the perfect home;" that is, a lawn.[4] Here is a sense of the importance of turfgrass today.[5]

- More than 1 million acres of farmland is devoted to grass seed and sodgrass cultivation.
- There are four million acres of managed turf in chronically water-short California
- In the Chesapeake Bay Watershed (CBW), turf comprises 5.3–9.5% of total watershed area.[6]
- In Maryland, 23% of the state's area within the CBW was turf in 2010.[7]
- Overall, there were about 128,000 square kilometers of managed turf in the United States as of 2012.[8]

**Question 14-2:** Calculate the acreage in hectares (ha) in managed turf in the U.S. based on 128,000 km$^2$. (There are 10,000 square meters/hectare).

The eastern United States was virtually cleared of virgin forest between 1850 and 1920 and has been partially reforested since 1920. Even so, less than half the forest cover of 1620 remains.

**Question 14-3:** In Virginia, lawn area increased from 714,000 acres in 1998 to 1,048,000 acres in 2004.[9] What percentage of Virginia (42,769 square miles, 27,372,160 acres) was turf as of 2004?

[2] Cooper S.R., and G.S. Brush. 1991. Long-term history of Chesapeake Bay anoxia. *Science, 254:* 992–996.
[3] See U.S. Energy Information Administration. 1996. *Emissions of greenhouse gases in the United States,* U.S. Dept. of Energy, Washington, D.C., p. 65, Figure 11. This publication includes maps showing the approximate extent of forest cover in the United States for 1620, 1850, 1920, and 1992.
[4] http://web.archive.org/web/20110125025810/http://www.edf.org//article.cfm?contentID=984.
[5] From the U.S. Department of Agriculture (USDA), World Resources Institute. 1997. *World Resources 1996–97; A Guide to the Global Environment* (New York: Oxford University Press); TPI Turf News 2005.
[6] Chesapeake Stormwater Network Tech. Bull #8, April 2010.
[7] Ibid.
[8] http://earthobservatory.nasa.gov/Features/Lawn/lawn2.php.
[9] www.vaturf.org.

**Question 14-4:** By 2011, 8.2% of that part of Virginia within the Chesapeake Bay Watershed (CBW) was turf. Roughly 60% of Virginia's area is within the CBW. Estimate how much of Virginia was turf in 2011.

**Question 14-5:** Assume the rate of increase from 1998 to 2004 to be constant. When would the entire state of Virginia be turf? Use $t = (1/r) \ln(N/N_0)$.

# WHAT IS TURF?

Turf is a grass *monoculture*. Even though there are over 5,000 species of grasses worldwide, sodded lawns contain very few species, and most warm-season grasses are dominated by a single species. Turf comes from one of two sources: (1) grass seed sown on a plot of soil, or (2) pre-grown turf rolls (also called sod) produced on turf "farms," (Figure 14-1, Table 14-1) which represent a significant land use in parts of the Pacific Northwest.

**FIGURE 14-1** Growing sod in the desert in Las Vegas, NV. (Courtesy of Robert Holmes/ Corbis Images.)

**TABLE 14-1** ■ Willamette Valley, Oregon Turf Acreage

| Year | Grass Seed Acreage | Sales |
|---|---|---|
| 1988 | 332,610 | $ 190 Million |
| 1997 | 410,510 | $ 300 Million |
| 2008 | 489,600 | $ 469 Million |
| 2010 | 375,665 | $ 228 Million |

Replacing turf with another crop is not an easy task in Oregon's Willamette Valley. Grass tolerates the extremely wet winters and very dry summers, but other crops do not do as well. Moreover, installing drainage systems in the Valley's volcanic, clay-rich soils to move rainwater off soaked fields in winter, necessary for most food crops, can cost $1,000 to $5,000 an acre.

**Question 14-6:**   What is the value in dollar sales per acre for each of the years?

Most new construction uses turf rolls from sod farms to give the house or commercial building an "instant lawn." According to the Professional Lawncare Network (PLN),[10] lawns do the following:

1. Produce oxygen as the plants photosynthesize: "625 square feet of lawn provides enough oxygen for one person for an entire day."
2. Cool the temperature: "On a block of eight average houses, front lawns have the cooling effect of 70 tons of air conditioning."
3. Control allergies by controlling dust and "replacing plants to which many people are allergic."
4. Absorb gaseous pollutants $CO_2$ and $SO_2$.
5. Trap particles (up to 12 million tonnes annually).
6. Protect water quality by filtering runoff.

**Question 14-7:**   Evaluate PLN claim 1, using critical thinking principles. Certainly turf produces more oxygen than dirt or asphalt, but how does turf compare with forest or other land uses? What additional information do you need to assess this claim?

**Question 14-8:**   Does claim 2 tell you what the cooling effect of turf is being compared to? Asphalt? Forest? Baked beans? Is the statement accurate? Precise? Clear? Ambiguous? What additional information do you need? Discuss.

---

[10]   www.landcarenetwork.org.

**Question 14-9:**   Analyze claim 3. Grass allergies are one of the most widespread and severe allergies in the United States. (See "For Further Thought" questions at the end of this Issue.)

## Adverse Effects of Lawns and Turf

Although sod is clearly better at erosion control than bare dirt, much turf has replaced forest, which is much more efficient at all the attributes described by the PLN. Sod has replaced agricultural land as well (see "For Further Thought" section for details).

Most lawns receive inappropriate doses of fertilizer, insecticide, and herbicide. For example, an annual nitrogen application of 3–4 lbs/1000 ft$^2$ will likely be necessary for new turf established on a site devoid of organic matter and plant nutrients, like most new suburban home sites. Improper fertilization of lawns can release substantial nitrate and phosphate into streams, fueling algal growth that leads to depletion of oxygen (hypoxia or anoxia). This may cause fish kills and contribute to infestations such as the *Pfiesteria piscicida* outbreak in 1997 that caused the closure of several streams tributary to Chesapeake Bay (Issue 21).[11]

Increases in turf area are associated with *sprawl development* (see Issues 24 and 25). Sprawl replaces agricultural land and woodland. Woodland is irreplaceable as wildlife habitat, is essential for aesthetics, and can help counter global warming induced by greenhouse gases.

"Caring" for turf has not-so-quietly become one of the nation's most polluting activities. Lawns require regular maintenance, according to the PLN. They should be watered when dry, mowed at regular intervals, and fertilized twice a year. Since most lawn-care devices (mowers, trimmers, blowers, weed whackers, and the like) are gas-powered, and many still have two-cycle engines, they burn fuel—a lot of it. In fact, a two-cycle gasoline-powered lawn mower run for an hour emits about the same amount of smog-forming emissions as forty new cars run for an hour, according to the State of California's Air Resources Board (ARB). Moreover, the World Health Organization (WHO) reports that the harmful effects of noise pollution are second only to polluted air in EU countries.[12]

Suburban lawns and gardens receive more pesticide applications per acre—3.2 to 9.8 pounds—than agriculture, which receives 2.7 pounds per acre on average. Of thirty commonly used lawn pesticides during the 2000s, nineteen were possible carcinogens, thirteen linked with birth defects, twenty one with reproductive effects, fifteen with neurotoxicity, twenty six with liver or kidney damage, and eleven have the potential to disrupt the endocrine (hormonal) system. The National Academy of Sciences estimates 50 percent of lifetime pesticide exposure occurs during the first five years of life.[13]

---

[11]   To assess the relative environmental impact of urban/suburban, pasture, cropland, and forested land uses, go to the EPA Chesapeake Bay Program website, http://www.epa.gov/Region3/chesapeake/ and http://www.chesapeakebay.net/.

[12]   *The Economist*, June 30 2012, p. 82.

[13]   www.beyondpesticides.org.

## The "Lawn Care" Industry

Increasing numbers of suburban households with working spouses find themselves with a decreasing interest in, or time for, lawn "maintenance." U.S. homeowners, including a growing number of retirees, are turning to lawn, landscape, and tree care professionals in record numbers. More than 29 million households regularly hired a "landscape service" by 2000. Cynics refer to it as one of the last "growth" industries in America. Lawn-care devices themselves, such as blowers, mowers, and trimmers ("weed whackers"), most of them gas-powered, represent a $5-billion-a-year growth industry. According to the Lawn Institute, Americans spent $6.4 billion caring for lawns in 2011.[14]

Is all of this cause for concern or congratulation?

**Question 14-10:** According to the Turf Producers International, a "well-maintained" 10,000 ft$^2$ lawn will generate around 1 ton (900 kg) of grass clippings annually.[15] On average, what is the weight of grass clippings (in pounds) produced per square foot of lawn per year? Restate your answer as kilograms per square meter.

**Question 14-11:** Based on this and the information in Questions 14-3 and 14-4, how many kilograms of grass clippings were produced from Virginia lawns in 2011?

Runoff from lawns has contributed to the widespread presence of pesticides in streams and groundwater. The chemical 2,4-D is the herbicide most frequently detected in streams and shallow groundwater from urban lawns.

**Question 14-12:** According to the city government of Mesa, AZ, lawns require 40–55 in. of rain or irrigation water per year, and Mesa gets around 8 in. of rain a year. Calculate the amount of water necessary to irrigate 5,000 ft$^2$ of lawn each year in Mesa, in addition to the rain. Express the amount in liters per day.

**Question 14-13:** How does this figure compare with the U.S. 2004 per capita water consumption of 1,800 gallons per day?[16] Convert your answer to liters.

---

[14] http://www.thelawninstitute.org/.
[15] www.turfgrasssod.org.
[16] U.S. Bureau of the Census, 2006. Statistical Abstract of the U.S.: 2006. www.census.gov/compendia/statab; www.waterfootprint.org/Reports/Hoekstra_and_Chapagain_2007.pdf

## ALLERGIES AND INDIVIDUAL RIGHTS

Many southwestern cities report increases in residents with allergies. In Tucson, "residents have twice the national rate of respiratory allergies."[17] Experts attribute the increases to people who move west from eastern states and bring a taste for lawns and eastern trees with them.

Some cities are beginning to ban certain plant species. Tucson has banned Bermuda grass (*Cynodon*), a key species in many lawns, and many western cities discourage turf plantings while encouraging "xeriscaping," that is, planting native drought-resistant plants in lawns to cut down on water demand.

However, some residents object that their rights are being violated. When Albuquerque's city council banned several types of trees, a resident responded, "I get pretty sick of people . . . saying that this is a desert and we have to live like it's a desert. We do not have to live like it's a desert. I love trees and I choose to live near them."[18]

**Question 14-14:** Do you think that city governments have an obligation, or the authority, to tell residents what they can or cannot plant? Do you think that people should have a right to plant whatever they like on their own property? Defend your reasoning.

**Question 14-15:** Health-care costs consistently have risen faster than the rate of inflation, and part of the cost of health care is treating people with severe allergies. Do you feel that those who insist on their right to plant whatever they choose on their property should be required to pay for the added burden on the health-care system that their actions cause or contribute to? How? Explain your reasons. Is this an environmental issue? Why or why not?

## GAS POWERED LEAF BLOWERS

"Leaf blowers: the most annoying . . . power tool ever invented."[19]

The gas-powered leaf blower was introduced to the United States as a lawn maintenance tool, but it gained little acceptance until drought conditions in California during the late 1970s led to restrictions on outdoor water use. In 2011, sales of gasoline-powered and electric leaf blowers exceeded 4.9 million units, at an estimated cost of $535 million.[20] A typical gas-powered blower uses about a third of a gallon of gasoline per hour of opera-

---

[17] Mendoza, M. Southwestern communities find greening desert is something to sneeze at. *Washington Post*, November 10, 1996.

[18] Ibid.

[19] www.noisefree.org.

[20] http://www.homechannelnews.com/article/leaf-blowers-numbers.

tion. By 2011, more than one hundred cities in California alone had banned or restricted leaf blowers. The California Air Resources Board advises citizens to avoid leaf blowers.[21]
Advantages and environmental disadvantages of leaf blowers are summarized below.

## Emissions from Leaf Blowers

Two-stroke engines power most gas leaf blowers. The two-stroke engine has several advantageous attributes. They are lightweight in comparison to the power they generate and operate in any position. Multipositional operation is made possible by mixing the lubricating oil with the fuel, which in turn leads to a major disadvantage: high exhaust emissions. As much as 30 percent of the fuel/oil mixture is exhausted unburned. Thus, exhaust emissions consist of both unburned fuel and products of incomplete combustion. The major pollutants from a two-stroke engine are oil-based particulates, a variety of hydrocarbons, and carbon monoxide.

Despite government-required reductions in emissions, in 2011, edmonds.com reported that consumer-grade leaf blowers still emit as much as 23 times the toxic carbon monoxide of a Ford F-150 truck. A two-stroke engine produces relatively little NOx.[22]

Some of these emissions are *toxic air contaminants,* as defined by the EPA. They include benzene, 1,3-butadiene, acetaldehyde, and formaldehyde.

Table 14-2 summarizes the ARB's best estimate of pollution from lawn-care devices, and, based on regulations effective in 2000, the contributions of leaf blowers to the state's air pollution in 2010. The ARB concluded, "lawn and landscape contractors, homeowners using a leaf blower, and those in the immediate vicinity of a leaf blower during and shortly after operation, are exposed to potentially high exhaust, fugitive dust, and noise emissions from leaf blowers on a routine basis." And, "the day-in-day-out exposure to [significant levels of] PM10 could result in serious, chronic health consequences in the long term." In 2009, ARB concluded that exposure to fine particulates, produced by diesel engines and devices like leaf blowers, were responsible for the "premature death" of 7,000–11,000 persons per year.[23]

**Question 14-16:** Should the health care "industry" in this country have to bear the burden of the effects of air pollution from leaf blowers on operators and bystanders? Justify your answer.

TABLE 14-2 ■ Inventory of Leaf Blower Exhaust Emissions (tons per day)

| Assessment of Two-stroke Engine Pollution | Leaf Blowers 2000 | Leaf Blowers 2010 | All Lawn & Garden, 2000 |
|---|---|---|---|
| Hydrocarbons, Reactive | 7.1 | 4.2 | 50.24 |
| Carbon Monoxide (CO) | 16.6 | 9.8 | 434.99 |
| Particulates (PM10) | 0.2 | 0.02 | 1.05 |
| (*Source:* California ARB) | | | |

[21] www.arb.ca.gov/msprog/leafblow/leafblow.htm.

[22] http://www.arb.ca.gov/html/brochure/pm10.htm.

[23] Ibid.

**Question 14-17:** Is it a proper role of government to monitor devices produced by corporations and to set standards for their operation, or should manufacturers be required to demonstrate that their products will cause no harm to the environment before they are allowed to produce and distribute these devices? Explain your answer.

### Effects of Resuspended Dust

Two additional sources of emissions from leaf blowers are "fugitive" (resuspended) dust and noise. We will consider dust next.

Leaf blowers are designed to move relatively large materials at hurricane-force wind speeds, and hence will also move much smaller particles, including those below 30 micrometers in diameter, which are not visible to the naked eye.

A Sebastopol CA environmental group reported that a leaf blower operated for one hour could suspend 5 pounds of dust, some of which could take days to resettle.[24]

An Orange County, California grand jury found that "fecal material, fertilizers, fungal spores, pesticides, herbicides, pollen, and other biological substances" were found in the dust resuspended by leaf blower use.[25] The ARB estimated that streets and sidewalks could contain about 3 grams of fine sediment per square meter, which would be blown into the air by a leaf blower. In addition, the chemical analysis of paved road dust showed small percentages of toxic metals arsenic, chromium, lead, and mercury, along with particles from tire and brake wear. Fine latex particles from tires are a known human allergen, but the effect of resuspending the material is unknown. Table 14-3 shows air pollution from leaf blowers compared to motor vehicles. Leaf blower users often blow dust and debris into the streets, leaving the dust to be resuspended by passing vehicles.

**Question 14-18:** The parking lots of the Potomac Mills Mall, Prince William County, Virginia, measure approximately 270,000 square meters. Based on an average value of 3 grams per square meter of dust, how much dust could be resuspended by "cleaning" the parking lots using leaf blowers daily? Report your answer in kilograms.

TABLE 14-3 ■ Air Pollution from Leaf Blowers Compared to Motor Vehicles (ARB report)

| | **Leaf Blower Exhaust Emissions, g/hr** | **Exhaust Emissions New Light-Duty Vehicle,*** |
|---|---|---|
| Hydrocarbons | 199.26 | 0.39 |
| Carbon Monoxide | 423.53 | 15.97 |
| Particulate Matter | 6.43 | 0.13 |
| Fugitive Dust | 48.6–1031 | N/A |
| * New light-duty vehicle represented vehicles one year old, 1999 or 2000 model year, driven for one hour at 30 mph. | | |

---

[24] www.progressivesource.org.
[25] www.arb.ca.gov/msprog/mailouts/msc0005/msc0005.doc.

# NOISE POLLUTION

We briefly referred to this earlier. In addition to damaging hearing, noise may cause other adverse health impacts. These include sleep disturbance, changes in performance and behavior, and acute annoyance. In fact, leaf blower noise may have contributed to at least one murder. According to the Bergen (NJ) *Record*, in May 2000 "a woman killed her 74-year-old neighbor by repeatedly running him over with a car, after their latest dispute, which involved his use of a leaf blower, police said."[26] The victim had complained to a neighbor earlier that the alleged assailant, wearing a dust mask, had threatened him earlier with a pitchfork due to the noise, emissions, and dust from the victim's leaf blower.

Long-duration, high-intensity sounds are the most damaging and usually perceived as the most annoying of all sounds. High-frequency sounds, up to the limit of hearing, tend to be more annoying and potentially more hazardous than low-frequency sounds.

Figure 14-2 shows some common noise levels and reference values from leaf blowers.

Based on extrapolations of EPA data, at least 3 million people nationwide could be routinely exposed to leaf blower noise at annoying levels. For California, the figure could exceed 300,000, based on the urban/rural population ratio and scaling of population values.

## Anti-Noise Legislation

The Federal Noise Control Act of 1972 sought "to promote an environment for all Americans free from noise that jeopardizes their public health and welfare." The EPA is charged with implementing this law.

About 13 percent of Californians live in cities that ban the use of leaf blowers, and six of the ten largest California cities, including Los Angeles, have ordinances that restrict

| Perceived Sound Level | Sound Level dB | Sound Level μPa | Examples | Leaf Blower Reference |
|---|---|---|---|---|
| PAINFULLY LOUD | 160 | $2 \times 10^9$ | fireworks at 3 feet | |
| | 150 | | jet at takeoff | |
| UNCOMFORTABLY LOUD | 140 | $2 \times 10^8$ | threshold of pain | Occupational Health and Safety Administration limit for impulse noise |
| | 130 | | power drill | |
| | 120 | $2 \times 10^7$ | thunder | |
| VERY LOUD | 110 | | auto horn at 1 meter | 90–100 dB leaf blower at operator's ear |
| | 100 | $2 \times 10^6$ | snowmobile | |
| MODERATELY LOUD | 90 | | diesel truck, food blender | 90 dB Occupational Health and Safety Administration permissible exposure limit |
| | 80 | $2 \times 10^5$ | garbage disposal | |
| | 70 | | vaccum cleaner | 82–75 dB leaf blower at 50 feet |
| QUIET | 60 | $2 \times 10^4$ | ordinary conversation | |
| | 50 | | average home | |
| VERY QUIET | 40 | $2 \times 10^3$ | library | |
| | 30 | | quiet conversation | |
| | 20 | $2 \times 10^2$ | soft whisper | |
| BARELY AUDIBLE | 10 | | rustling leaves | |
| | 0 | $2 \times 10^1$ | threshold of hearing | dB = decibels μPa = micro Pascals |

**FIGURE 14-2** Noise levels from common sources and some leaf blower comparisons. (Source: California ARB)

---

26 www.NorthJersey.com.

or ban leaf blowers. All together, to summarize, about one hundred California cities have ordinances that restrict either leaf blowers specifically or all gardening equipment generally, including cities with bans on leaf blower use.

**Question 14-19:** Approximately 14 million new motor vehicles are sold in the United States each year, each of which contains a catalytic converter that costs approximately $300. Calculate the total amount Americans pay for catalytic converters to take pollution out of the air, and then comment on whether it makes economic sense to put that pollution back in the air in the form of toxics from leaf blowers. Defend your answer.

**Question 14-20:** Summarize the positive and negative attributes of millions of acres of American lawns. Be sure to justify or include evidence for your choices. In what ways might lawns be compatible with the principles of sustainable communities?

**Question 14-21:** Summarize the main points of this Issue.

# FOR FURTHER THOUGHT

**Question 14-22:** Go to www.noisefree.org, and read representative stories on leaf blowers. Summarize in a couple of paragraphs what you found.

**Question 14-23:** Contact your city or county department of waste management. Do they accept lawn debris or waste? Do you think this is an appropriate use of public funds?

**Question 14-24:** Go to EPA's website www.epa.gov. Find out how serious emissions from gasoline-powered lawn-care devices are, and what regulations are in effect or proposed to control emissions from gasoline-powered lawn-care devices.

**Question 14-25:** Contrast the impact of gasoline-powered with electric lawn-care devices. What advantages do electric devices provide? What disadvantages?

**FIGURE 14-3**   The Reserve Golf Course in Pawleys Island, SC. This course was designed to be environmentally sensitive and has much less turf than a typical golf course. (Courtesy of The Reserve Golf Club, McConnell Golf, LLC.)

**Question 14-26:**   Golf courses typically replace fields or forests with turf. Some newer courses (Figure 14-3) are reducing the area of turf. Research this issue. Check the golf courses in your area. Are they constructed and maintained in a sustainable way?

# Issue 15

MOUNTAINS OF TRASH:
*ARE THEY SUSTAINABLE?*

## KEY QUESTIONS

- What specific issues are involved with construction and demolition (C&D) waste?
- How much municipal solid waste (MSW) do Americans generate every year?
- What happens to this MSW?
- How much waste is transported across political boundaries for disposal?
- How does this practice rely on the U.S. Constitution?
- What is e-waste, and why is it important?
- To what extent is waste consistent with sustainability?

## INTRODUCTION: CONSTRUCTION AND DEMOLITION WASTE (C&D)

In May 2012, Horry County, SC, prepared to open its spanking new construction and demolition (C&D) waste handling facility. A citizen wrote in support of collecting the waste, but shipping it "away." He said, "I believe we should keep our area free from overflowing landfills, that if we keep all the garbage here we may be looking at a [300 foot high] Mount Horry."[1]

In New England, C&D waste handling facilities proliferate, and much of it is from out of state. As of 2006, at least fifteen Maine communities had C&D incinerators or dump sites. Much of the waste is burned. Local activists argue that Maine, as a result, has the highest asthma rate in the United States. The state has issued advisories on the dangers of eating fish due to dioxin contamination. And Somerset County is rated among the worst 20 percent of counties in the United States for air releases of identified carcinogens. As of 2006, the Northeastern states (New England, NY and NJ) generated about 12 million tons of C&D waste, nearly all of it "landfilled" or incinerated.[2]

What is C&D waste? EPA says C&D waste is "the debris generated during the construction, renovation, and demolition of buildings, roads, and bridges, and often contain bulky, heavy materials, such as concrete, wood, metals, glass, and salvaged building components." EPA further estimates that 136 million tons of C&D waste is generated each year.

Question 15-1: What is the per capita C&D waste per year in the U.S.? The U.S. population is about 310 million.

---

[1] www.myrtlebeachonline.com.
[2] http://www.epa.gov/region1/solidwaste/cnd/.

144

## Focus: Pressure-Treated Wood and C&D Waste[3]

Why is C&D waste controversial? One reason is *pressure-treated* wood. Chromated copper arsenate (CCA) is a wood preservative containing chromium, copper, and arsenic. CCA is used to protect wood from attack by insects and microbes. As you might imagine, CCA is a very toxic substance: inorganic arsenic is a known human carcinogen, for example. The average amount of arsenic in soils below decks made with CCA-treated wood increases with age, and it can easily reach seven to eight times the maximum safe concentration.[4]

Since the 1970s, the majority of the wood used in outdoor residential construction (decks, patios, child's playsets, etc.) has been CCA-treated wood. Effective December 31, 2003, CCA was banned for residential uses, with certain exceptions, but vast quantities have been used to treat wood for construction over the past sixty years.

C&D is just one category of waste. Other categories are industrial waste, toxic waste, sewage sludge, electronics waste (e-waste), and MSW. In this Issue, we will focus on MSW and discuss the specific issue of shipment of MSW across political boundaries for disposal. However, MSW is only a small category of waste in the United States. One source estimates it to represent only 2 to 20 percent of the country's total annual waste.[5] Lastly, we will briefly introduce a rapidly growing waste category: e-waste.

## Municipal Solid Waste (MSW)

Figures 15-1a and 15-1b show, respectively, categories of MSW and mass of MSW generated from 1960 to 2010.

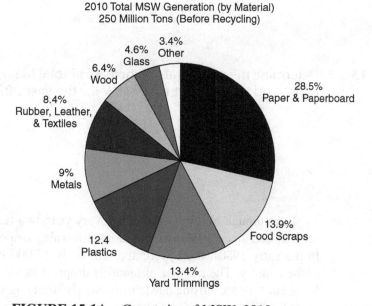

**FIGURE 15-1A**   Categories of MSW, 2010.

---

[3] www.epa.gov/oppad001/reregistration/cca/.

[4] www.caes.state.ct.us/PlantScienceDay/1999PSD/arsenic99.htm.

[5] www.zerowasteamerica.org.

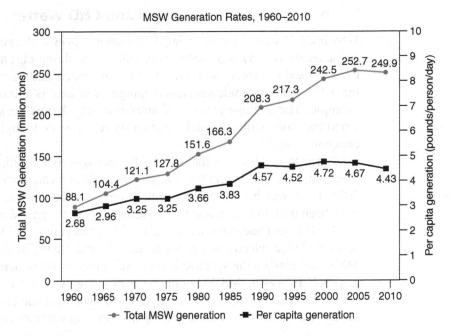

**FIGURE 15-1B**    Trends in MSW, 1960–2010

**Question 15-2:**    By what factor is 2010 MSW greater than 1960 MSW?

**Question 15-3:**    Determine the annual rate of increase of total MSW, based on 1990 and 2010 total MSW, and then project total MSW for the year 2020. Use $r = (1/t) \ln(N/N_0)$.

Even though MSW is increasing every year, two-thirds of the nation's MSW landfills have closed since 1980 as regulations governing disposal of MSW radically tightened.[6] In the early 1980s, there were approximately 13,000 municipal waste dumps (landfills) in the country. The number of landfills dropped to 8,000 by 1988 and to 1,750 by 2008. More landfills are closing due to increasingly strict regulations. Industry sources report that some landfill operators have reduced dumping fees to fill as much landfill space as possible before new regulations go into effect.[7]

During the 1990s, in most areas of the country, the cost of MSW disposal leveled off or even declined, mainly due to the construction of giant regional landfills that charge

---

[6] For details on landfill regulations, go to www.epa.gov/epawaste/index.htm.
[7] www.wasteage.com.

relatively low fees for disposal. So much additional capacity has been added that there is now an excess of disposal capacity in many areas.

From 1986 to 1996, construction of incinerators ("waste-to-energy facilities") mushroomed, and recycling grew as well. According to the Congressional Research Service (CRS), by 1995 incinerators were burning 16 percent of the nation's MSW.[8] But incinerators have been increasingly shunned due to high construction and operating costs, as well as environmental concerns about air quality. Moreover, incinerators leave a residue of ash, sometimes containing toxic substances, that must be buried somewhere.

Landfills and incinerators are of course not the only answer to waste disposal. According to the EPA, by 2010 approximately 9,000 local governments had curbside collection of recyclable materials and around 3,090 operated composting programs for yard waste.

**Question 15-4:** Thirteen percent of MSW was "yard waste" by 2010. Should it be the government's job to dispose of "yard waste"? Why or why not?

Figure 15-2 shows changes in recycling (including composting) rates to 2010.

**Question 15-5:** How has the recycled percentage of MSW changed since 1990 in the United States?

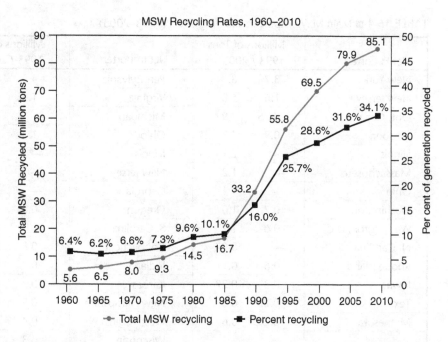

**FIGURE 15-2** Changes in recycling (including composting) rates to 2001.

---

[8] McCarthy, J. E. Congressional Research Service Report for Congress Interstate Shipment of Municipal Solid Waste: 2004 Update.

In Europe, waste charges are generally assessed by weight, varying between $0.45 and $0.80 a kilogram.

**Question 15-6:** Identify advantages and disadvantages to a waste-disposal charge based on weight.

# TRUCKIN' TRASH

Today, it is commonplace for waste from New York City to be transported up to 1,000 kilometers for disposal in huge commercial landfills in Pennsylvania, Virginia, and other states; for industrial waste from Michigan to be hauled to the Dakotas for dumping; for MSW from Toronto, Ontario, Canada, to be shipped to Michigan (770,000+ tons/yr); and for waste from Los Angeles to be transported by truck and rail into the Mojave Desert.

The CRS reports that of 236 million tons of MSW generated in 2003 (the last year for which CRS had data), at least 39 million tons were transported across state or national boundaries for dumping.[9]

Table 15-1 shows exporting and importing states, as of 2004.

**Question 15-7:** What percentage of MSW was shipped out of state in 2003? This compares to 9 percent in 1997.

**TABLE 15-1** ■ Main MSW Exporters and Importers (1993–2003)

| Net Exporters | Millions of Tons 1993 / 2003 | | Net Importers | Millions of Tons 1993 / 2003 | |
|---|---|---|---|---|---|
| New York | 3.7 | 8.2 | Pennsylvania | 4.8 | 9.2 |
| New Jersey | 1.6 | 5.8 | Virginia | 1.5 | 5.5 |
| Ont, Canada | 0.25 | 2.9 | Michigan | — | 4.5 |
| Missouri | 0.9 | 2.3 | Ohio | 1.3 | 2.5 |
| Illinois | — | 2.1 | Illinois | — | 1.9 |
| Massachusetts | — | 1.2 | New Jersey | — | 1.7 |
| Ohio | — | 1.1 | Georgia | — | 1.45 |
| Washington | 0.7 | 1.0 | Oregon | 0.8 | 1.4 |
| Washington, D.C. | 0.6 | 1.2 | S. Carolina | — | 1.2 |
| N. Carolina | — | 0.97 | Connecticut | 0.8 | 0.05 |
| Rhode Island | 0.6 | 0.11 | Indiana | 0.7 | 0.9 |
| Iowa | 0.3 | 0.27 | Kansas | 0.7 | 0.7 |
| Texas | 0.23 | 0.5 | New Hampshire | 0.5 | 0.4 |
| Minnesota | 0.2 | 0.6 | West Virginia | 0.4 | 0.28 |
| | | | Wisconsin | 0.3 | 1.2 |

*Source:* McCarthy, J.E. 2004. *Interstate Shipment of Municipal Solid Waste: 2004 Update.* Congressional Research Service report RL32570

[9] http://www.epa.gov/epawaste/nonhaz/municipal/images/index_msw_recycling_rates_900px.jpg.

In 1997, cost of diesel fuel was around $1.10 per gallon, and transport of waste cost around $1 per mile per ton. In 2012 diesel was around $4 per gallon. In 1997, 18 million tons of MSW was shipped out of state. Transportation costs alone (not including disposal) amounted to over $500 million.

**Question 15-8:** Calculate the average transport cost per ton that municipalities were willing to pay in 1997 to ship worthless garbage across state lines, rather than deal with the waste in their own locality.

**Question 15-9:** Estimate the order of magnitude of change of transport cost, based on $4 per gallon for diesel fuel.

As you have seen, out-of-state shipment of trash increased significantly during the period 1993–2003, especially for those states on the eastern seaboard. To cite one example, Maine pays about $600 per ton to ship biomedical waste out of state, although in-state disposal is cheaper.

## The Closing of Fresh Kills Landfill

In 2001, Staten Island's Fresh Kills landfill closed, and 13,000 tons of garbage a day, almost 5 million tons a year, found a new home. Montana's Senator Max Baucus described it this way in a speech on the Senate floor in 1997:

> ". . . about 1200 trucks of garbage a day coming out of New York City . . . a convoy of trash trucks 12 miles long, 365 days a year. . . . The entire state of New York can take only about 1200 tons of New York City's trash each day and that means the rest, over 4 million tons a year, must go out of state."[10]

New York City has among the most stringent recycling regulations in the country; however, many things are not recycled. Table 15-2 shows items that are not recycled in New York City.

**TABLE 15-2 ■ Items not recycled in New York City**

| Household Items | May be placed on curb |
| --- | --- |
| Paper | Soiled paper cups and plates, photographic paper, plastic-or-wax-coated paper, napkins, paper towels, tissues, hardcover books |
| Plastic | Styrofoam, cups, bags, wrap |
| (Source: www.nyc.gov/html/nycwasteless/html/recycling/recycling_nyc.shtml) | |

---

[10] www.baucus.senate.gov.

**Question 15-10:**  Calculate the amount of diesel fuel needed to truck a 15-ton load of MSW from New York to Virginia, a distance of about 600 km (400 miles). Fifty-three foot diesel-powered trucks get about 4 mpg. Remember the truck has to return (empty) for another load.

**Question 15-11:**  Since the truck has to return empty, how much of the fuel is essentially wasted? Diesel exhaust is one of the most severe types of air pollution.

**Question 15-12:**  Study Table 15-2. Suggest in a paragraph or two ways that New York City's MSW can be reduced.

# SUPREME COURT RULINGS ON GARBAGE

Federal court rulings have had a profound impact on local waste management programs. The Supreme Court has repeatedly ruled that, based on the Constitution's Commerce Clause, Congress has sole authority to regulate interstate commerce. The Commerce Clause, Article 1, section 8 (3) states, "The Congress shall have the power to regulate commerce with foreign nations and among the several States." Since the late 1980s, Congress has considered, but never enacted, bills that would allow states to impose restrictions on interstate waste shipments. Without congressional authorization, states do not have the right to ban waste shipments.

In three decisions since 1992, the U.S. Supreme Court has ruled that shipments of waste, even though "valueless" in the Court's words, are acts of "commerce" and thus protected under the U.S. Constitution from state or local laws banning the importation of municipal garbage. As a result, state and local governments may not prohibit private landfills from accepting waste from-out-of state nor may they impose fees that discriminate on the basis of origin on waste disposal from another state.

In *City of Philadelphia v. New Jersey,* a landmark 1978 case, the U.S. Supreme Court ruled seven-to-two that New Jersey could not bar MSW from Philadelphia (437 U.S. 617):

> New Jersey sought to impose upon out-of-state commercial interests the full burden of conserving the state's remaining landfill space. . . . The New Jersey statute cannot be likened to a quarantine law that bars articles of commerce because of their innate harmfulness, and not because

of their origin. New Jersey concedes that out-of-state waste is no different from domestic waste but it has banned the former. . . . Today cities in Pennsylvania and New York find it expedient or necessary to send their waste into New Jersey for disposal Á tomorrow New Jersey may find it expedient or necessary to send their waste into Pennsylvania or New York for disposal and then these states might claim the right to close their borders.

**Question 15-13:** Do you believe the Supreme Court acted correctly in ruling that garbage is "commerce" even though it is valueless? Explain.

## TRUCKING TRASH AND HIGHWAY SAFETY

Although trash hauling has been a bonanza for some railroads and barge companies, most trash is hauled by truck. Citizens across the country are becoming increasingly concerned about the safety of truck waste transport as well as the sheer numbers of trash trucks. Officials randomly inspect trash trucks and have documented a history of safety violations, including the following examples:

- During a five-day "crackdown" in May 2006, Pennsylvania state police inspected 3,968 trash trucks and found 469 violations on 368 trucks. Ten years earlier, a similar crackdown cited 192 of 849 trash trucks with one or more violations.[11]
- An inspection of seventy-two trash trucks on December 12, 2005, on I-84 in Pennsylvania cited nine with one or more violations.[12]
- According to the *New York Times,* 9,564 large trucks were inspected on Long Island in 2005, and 35 percent had violations serious enough for them to be taken out of service.[13]

**Question 15-14:** To what extent are victims of trash-truck crashes part of the real cost of waste disposal (i.e., that part of the cost not reflected in the price paid to landfill operators, etc.)? In your answer discuss whether you think this is a "fair" question and state your reasons.

---

[11] PA Dept of Env Protection, http://www.portal.state.pa.us/portal/server.pt/community/newsroom/14287?id=6493&typeid=2.

[12] PA Dept Env Protection.

[13] *New York Times,* http://select.nytimes.com/gst/abstract.html?res=F00E12FB3B540C7A8CDDAD0894DE404482.

# WASTE DISPOSAL COSTS: WHAT ARE THE ALTERNATIVES?

The cost of MSW disposal varies widely. For one example, consider the Tehama County Sanitary Waste Facility at Red Bluff, CA. It charged $44 per ton for MSW, $40 per ton for yard waste and metal, nothing for clean C&D waste, and $6 for each mattress or box spring.[14]

The alternatives to landfilling, incineration, and recycling waste are *waste reduction* and *reuse*. Rather than throw "away" appliances, we could require that they be disassembled and reused. Rather than discard hi-tech equipment like televisions, DVD players, cell phones, and increasingly, cars, they could be made of reusable components.

Rather than throw food waste into trash, much could be composted. (Meat wastes are not recommended for composting, however.) The European Union is presently formulating regulations that require manufacturers to take back items when the consumer is finished with them. The EU is considering banning landfilling of waste by 2025.

# E-WASTE: THE TWENTY-FIRST CENTURY'S MOST PRESSING WASTE ISSUE?

*The Ecologist* has called electronic waste "the world's fastest growing and potentially most dangerous waste problem."[15] E-waste—cell phones, televisions, computers, video games, VCRs, DVD players, fax and copy machines, etc.—is the fastest growing category of waste in the United States. Various reports put electronic waste at less than 5 percent of the total solid-waste stream, but according to the EPA e-waste is growing two to three times faster than any other category. Most people have no concept of the sheer volume and environmental impacts of "high-tech trash." Do you?

Some e-waste facts:[16]

- In 2007, Americans disposed of 140 million cell phones—only 10% were recycled.
- Again in 2007, we disposed of 205 million computer-related items (monitors, printers, keyboards, copiers, etc.) of which 18% were recycled.
- In 2009, we bought 438 million new "consumer electronics."
- In 2008, we discarded 3.16 million tons of e-waste, only 430,000 tons were recycled, with 2 million tons of e-waste going to U.S. landfills (EPA). But two thirds of what's not being used, trashed, or recycled is simply "stored" at home or work.
- Nearly 80 percent of U.S.-generated electronic waste is "exported" to Asia and Africa (Figure 15-3), where impoverished people scavenge the waste for saleable materials, taking few precautions. Such waste contains toxic materials and may produce dioxin when burned.
- Television monitors can contain as much as 8 pounds (3.6 kg) of lead. Replacing cathode-ray tubes (CRTs) with flat-panel displays eliminates the need for lead shielding.
- Only 17% of computers discarded in 2009 were recycled.
- Virtually every chip manufacturer from the 1970s and 1980s is responsible for a Superfund site (from leaking underground storage tanks, etc.).
- Heavy metals like lead and mercury and toxics like flame retardants are contained in high-tech products. The extent to which they can leak into landfills has been little studied.
- Chip maker Intel reports that it takes 16 gallons of water to make one microchip— far less than in 1998.

---

[14] For a complete list, visit http://www.co.tehama.ca.us/index.php?option=com_content&view=article&id=124&Itemid=253.

[15] *The Ecologist*, June 2009.

[16] http://www.electronicstakeback.com/wp-content/uploads/Facts_and_Figures, and epa.gov.

**FIGURE 15-3A**
"Recycling" at
the Agbogbloshie
Market in Accra,
Ghana. (D. Abel)

**FIGURE 15-3B**
Extracting metals
from computer
components over an
open fire. (D. Abel)

**FIGURE 15-3C**
Children search the
dump around the
market for recyclable
items. (D. Abel)

- There are hundreds of pending legal cases of birth defects that may be related to chip manufacturing.
- In the United States, it can be difficult to recycle e-waste. In Japan, a resident can recycle e-waste at the Post Office. There, legislation has been instrumental in e-waste recovery—up to 70 percent was recovered in 2005.

**Question 15-15:** Canada's computer industry believes consumers should pay to get rid of their e-waste. The industry wants to add a $25 fee to the cost of computers to cover recycling. Do you support such a plan? Why or why not?

**Question 15-16:** It has been said that in Europe "anything with a plug is recycled." Would you support this practice being required by law in the United States and Canada? Why or why not?

## E-Recycling

You can access http://www.electronicsrecycling.org/Public/default.aspx, the National Center for Electronics Recycling, for information on high-tech recycling. Many electronics manufacturers—Apple, Dell, IBM, Xerox, Canon USA, Gateway, Hewlett-Packard, and Nokia—already provide some type of recycling program. Customers usually pay a fee or shipping charges. Contact the electronics manufacturer or retailer for more information.

**Question 15-17:** Summarize the main points of this Issue.

**Question 15-18:** Discuss MSW and e-waste from the perspective of sustainability.

# FOR FURTHER THOUGHT

**Question 15-19:** Ontario, Canada, ships much of its MSW to Michigan for disposal. What are the advantages and disadvantages to Michigan of this activity? Is this a "national security" issue? Why or why not?

**Question 15-20:** Go to the website of Zero Waste America (www.zerowasteamerica.org) and analyze "Three Steps to Zero Waste in the U.S."

1. What are the three steps? Which is least controversial in your view? Why?

2. Assess the likelihood of each of the steps being implemented. You could use a key such as, "most likely," "very likely," "possible, but unlikely," and "unlikely." Justify your conclusions.

**Question 15-21:** How does your city, county, or school handle yard waste?

**Question 15-22:** www.recycle-steel.org reports that the U.S. steel-recycling rate hit an all-time high of 75.7 percent in 2005, recycling 76 million tons of steel. In 2010, 68 percent of cans were recycled. What factors discourage can recycling? Can it eventually approach 100 percent?

**Question 15-23:** Waste Management (WM), the nation's largest waste hauler, operates a fleet of more than 22,000 trucks. In July 2011, it added its 1,000th truck fueled by landfill gas (methane) obtained from WM landfills. Research "landfill gas." What trends in methane extraction and use can you identify?

**Question 15-24:** Search the Internet for the article *How Europe's Discarded Computers Are Poisoning Africa's Kids.*[17] What is your reaction to the article?

---

[17] Available at: http://www.spiegel.de/international/world/the-children-of-sodom-and-gomorrah-how-europe-s-discarded-computers-are-poisoning-africa-s-kids-a-665061.html

# Issue 16

## GOLD MINING:
## *IS IT SUSTAINABLE?*

### KEY QUESTIONS

- What is a mineral deposit?
- How is gold mined?
- What are the environmental impacts of gold mining?
- Could gold recycling satisfy world demand?

### INTRODUCTION

#### What Is a Mineral Deposit?

"The single most polluting activity done on Earth is mining for gold."[1]

The term "mineral deposit" refers to economically valuable mineral assemblages that can be extracted and refined at a profit. With rare exceptions, mineral deposits must have undergone *enrichment* by some natural process, before they are useful to humans. Iron, for example, makes up nearly 6 percent of the Earth's crust, but it must be concentrated up to about 50 percent to be economically viable.

The concentration at which deposits become economically valuable and are then called *ores* depends upon such variables as the world price of the commodity; the efficiency of processing to extract the valuable element(s) from the ore; any *subsidies* provided by government such as tax incentives, low-interest loans, or lax enforcement of environmental laws; shipping costs; currency fluctuations; and the costs of mitigating (that is, reducing) environmental impacts.

**Question 16-1:** Assume you were a government official in charge of economic development. List a few types of subsidies that you might propose to encourage mining, and explain your choices.

---

[1] http://academic.emporia.edu/abersusa/go336/morris/.

156

## How Do We Use Minerals?

Human societies have exploited mineral deposits for thousands of years. Attempts to determine the span of time over which the Greek epic poems *The Iliad* and *The Odyssey* were written focused on the poet's mention of iron and *bronze*—an alloy of copper and tin—in weaponry.

The U.S. Geological Survey publishes an annual *Minerals Yearbook,* tracking the production and use of nearly ninety mineral commodities from "abrasives" to "zirconium." You can access information on minerals there.[2]

During the past century and a half, *steel* has been manufactured by *alloying* various elements with iron to provide the characteristics exhibited by steel varieties. *Stainless steel,* for example, contains a thin surface layer of nickel and/or chromium. Cobalt is used to impart heat resistance to steel products, especially the engines of high-speed jet aircraft and rockets. The weapons of modern warfare have placed great demands on such relatively rare metals as cobalt, chromium, and molybdenum. High-tech devices like "smart phones" require rare Earth metals.

Demand for metals for weaponry means that many nations, like the United States, are dependent on other nations for their entire supply of these commodities. The uses to which we put all of these materials are beyond the scope of this Issue. We discussed coal mining in Issue 8. Here we will focus on gold.

# GOLD AND GOLD MINING

For millennia, humans have lusted after precious metals. Gold was behind the invasion of the New World by Spain in the sixteenth century, and gold and silver looted from Native Americans financed Spain's wars during that era. Precious metals stolen from Spanish galleons by privateers chartered by Queen Elizabeth I made a sizeable addition to the English treasury. So much gold and silver was brought back from the Americas that it set off an extraordinary period of inflation and may have contributed to the collapse of Spain as a great power—a classic example of too much money chasing too few goods.

Half of the gold ever produced on Earth has been produced since 1960. Figure 16-1 shows world cumulative gold production.

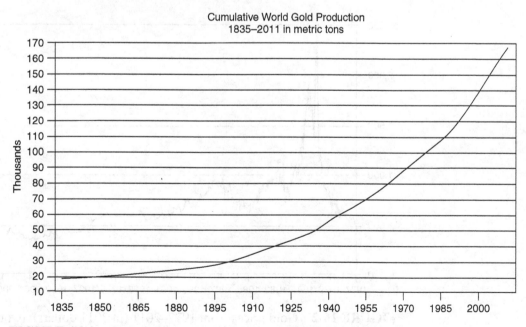

**FIGURE 16-1**   Cumulative gold production, 1835–2011 (tonnes).

---

[2] http://minerals.usgs.gov/minerals/.

**Question 16-2:** Estimate the total gold production during the 20th century.

The World Gold Council[3] reports that the electronics industry is the world's third biggest user of gold, after jewelry and coins. Gold-plating of contacts accounts for most of the gold used by the electronics industry—over 200 tonnes (6.5 million oz) annually. Gold potassium cyanide (GPC), or *plating salts,* accounts for about 70 percent of all electronics gold use. Contacts are normally electroplated with a very thin film of GPC, 2.5 microns or less.

**Question 16-3:** What is your "reaction" to the chemical called gold potassium cyanide?

Figure 16-2 shows the price of gold since 1977.

**Question 16-4:** What is the world price today? A good place to look is bloomberg.com.

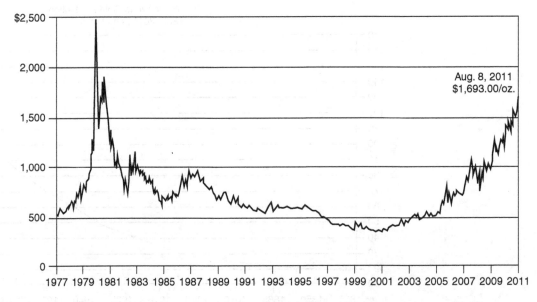

**FIGURE 16-2** Gold prices from 1977–2011 (in 2011 dollars). (Courtesy of NPR.)

---

[3] www.gold.org.

The high price of gold is forcing the electronics industry to use gold more sparingly, as well as to evaluate alternatives. Electronics manufacturers, however, have as yet found no practical substitute for gold. As a result of its use in minute quantities in billions of pieces of electronics equipment, gold is increasingly becoming incorporated into e-waste, which we discuss in Issues 15 and 27.

# MODERN METHODS OF GOLD MINING

Beginning in the second half of the nineteenth century, industrial-scale gold mining used high-pressure hoses to pry gold-bearing veins free from igneous rocks in the Sierra Nevada of California. This activity unalterably changed rivers throughout the northern Sierra by depositing millions of tons of sediment in river channels over the course of a few decades.

Today, the left-over cost of mining remediation is left to the State of California and federal agencies. Gold mining represents an economic activity that flourished because part of the cost of the activity—environmental degradation—was not included in the price. Gold mining continues to have a destructive environmental impact in regions where the "cyanide-leach" and "amalgamation" methods of gold recovery are practiced today.

The National Mining Association[4] provides information about toxic releases from mining and materials processing. The following relate to gold mining:

*Cyanide (CN).*   Cyanide is widely used in gold mining. It is one of the rare reagents that will dissolve gold in oxygenated water. Cyanide concentrations used in gold ore-processing solutions range from 0.01 percent to 0.05 percent.

**Question 16-5:**   Express these concentrations in parts per million or ppm (10,000 ppm = 1%).

Cyanide is slowly degraded by oxidation and is not geologically persistent in the environment. While cyanide is extremely toxic, it does not *bioaccumulate* and is not carcinogenic or mutagenic. However, both sodium cyanide (NaCN) and potassium cyanide (KCN) are extremely deadly compounds. And both are widely used in gold mining.

*Mercury (Hg).*   Mercury used in gold mining is easily released into the environment where it can remain indefinitely. Some portion of inorganic mercury will be slowly transformed into a form of organic mercury called *methylmercury* by bacteria in soil or water. Methylmercury is extremely toxic.

## Gold Mining by Cyanide Leaching[5]

Cyanide "heap leaching" uses cyanide solutions to recover gold from huge piles of crushed low-grade ore. Leaching occurs in "leach heaps," using drip emitters or sprays to distribute the cyanide solution.

The process set off a new U.S. "gold rush" in the late 1980s and early 1990s by making it profitable to mine rock containing tiny amounts of metal.[6] It works like this. Rocks containing gold are crushed to increase surface area. Since metals are present in "host" rocks in very small quantities, millions of tons of rock must be processed to yield profitable quantities of gold.

---

[4] http://www.nma.org/statistics/gold_silver.asp.
[5] www.mine-engineer.com.
[6] www.fws.gov.

In general, the finer the gold, the quicker it will dissolve in an alkaline-oxygenated solution of sodium cyanide (NaCN). A 45-micron particle of gold would dissolve in about half a day, while a 150-micron particle might take up to two days. The use of alkali solutions is essential: in an acid pH, the cyanide solution would decompose to produce deadly hydrogen cyanide (HCN).

The rate of dissolution of gold is directly proportional to oxygen concentration as well: no oxygen, no dissolved gold. The presence of other elements and compounds such as copper, zinc, silver, and especially sulfides complicate the process, and one or more of these is almost always present in gold ores. The presence of sulfides, for example, reduces gold dissolution and makes the process uneconomic in most cases.

Heap-leach gold mines collect cyanide-laced wastewater in huge holding ponds, some of which can cover 60 acres. Figure 16-3 shows a "tailings pond" collecting cyanide wastes from a heap pile, during gold mining near Elko, Nevada. If these ponds fail, an ecological disaster will likely result. And that is just what happened in Romania on January 30, 2000. It has been called Europe's worst natural disaster after Chernobyl (1986).

***Focus: The Baia Mare, Romania, cyanide disaster.[7]*** The area around Baia Mare, Romania, has been the site of mining activity for centuries. (Figure 16-4 shows the locality, in northwest Romania.) At the time of the disaster, there were at least 215 tailings ponds storing toxic mining waste.

The region has severely polluted soils and rivers from mining and industrial activity, the latter carried out under the former communist regime. The region's Sasar River is also known as the "Dead River." The World Health Organization (WHO) describes the area as a toxic "hot spot." Many residents live within 50 meters of toxic discharges. Blood lead levels in the local population are among the "highest ever recorded," according to the WHO. In some children, blood lead concentrations are six times the maximum "safe" levels. The lead is from the operation of lead smelters in the area, using little or no pollution control.

**FIGURE 16-3** A "tailings pond" collecting cyanide wastes from a heap pile during gold mining near Elko, Nevoda. (Courtesy of U.S. Fish and Wildlife Service.)

---

[7] www.rec.org.

The mining company, Aurul, jointly owned by an Australian concern and a Romanian partner, obtained necessary environmental permits in 1999 and began operations to reclaim or "mine" the numerous abandoned tailings piles in the region. Aurul's plan was to use higher than usual cyanide concentrations to leach precious metals from the tailings.

Following heavy rains and snow melt, at 10 P.M. on January 30, 2000, cyanide-laced wastewater further contaminated with heavy metals spilled from a collapsed tailings dam encircling a waste pond at the Aurul plant site. The spill discharged 100,000 m$^3$ of poisoned water into adjacent drainages, eventually affecting 2,000 kilometers of the Danube River's drainage system. Hungary alone reported at least 1,200 tonnes of dead fish.

The waste contained 50 to 100 tonnes of cyanide, as well as heavy metals, primarily arsenic, cadmium, copper, and lead. The discharge eventually reached the Danube itself, and finally the Black Sea. At the Danube Delta 1,000 kilometer downstream, 29 days after the spill, cyanide concentrations measured 58 micrograms per liter. For comparison, the Rhine River standard for maximum cyanide concentration is 25 micrograms per liter.

**Question 16-6:** Express 100,000 m$^3$ in liters and gallons.

**Question 16-7:** Based on the above data, what was the average concentration of cyanide, in mg/L, in the original 100,000 cubic meters of poisoned water?

The UN Environment Programme (UNEP Mineral Resources Forum) summarized the impacts of the spill as follows[8]:

- Contamination of drinking water of 2.5 million people at twenty-four locations.
- Thousands of tonnes of dead fish, as well as destruction of other aquatic species in the river systems. Fish are 1,000 times more sensitive to cyanide than humans.
- Severe negative impact on biodiversity, the rivers' ecosystems, and socioeconomic conditions of the local population. Heavy metals, including lead and copper, bioaccumulate. Copper is especially soluble in water and very toxic to aquatic plants.

UNEP further found that, "There were no provisions, such as water pumps, for coping with . . . a rise of pond water level due to uncontrollable input into the reservoir system. [Heavy rain and melting snow] aggravated the situation leading to an uncontrolled rise of pond level resulting in the overflow of the dam."[9]

UNEP summarized factors contributing to the accident as the following:

1. Deficiencies in design of the tailings retention dam and cyanide-treatment processes at the mine, especially regarding emergency measures to be implemented under "unusual operating conditions"
2. Deficiencies in planning and precautions against spills and poor or nonexistent emergency response plans
3. Inappropriate permitting of the facility on the part of government agencies
4. Inadequate monitoring and inspection by Romanian authorities

---

[8] Available at http://www.scribd.com/doc/97207138/Bp-Em-Cyanide.
[9] Ibid.

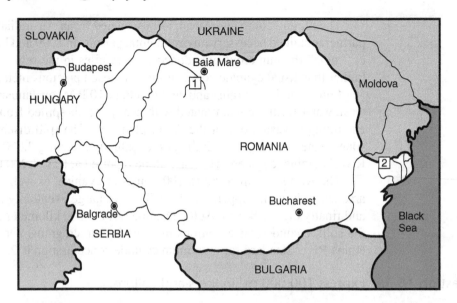

**FIGURE 16-4**   Location of Baia Mare in northwest Romania and path of the cyanide pollution from the Aurul spill, (1) January 30, spill at site; (2) February 28, spill reaches Danube Delta and Black Sea.

Path of the pollution plume is shown in Figure 16-4.

**Question 16-8:**   The UN Assessment recommended, among other things, "an international system for addressing the issues of liability and compensation related to such spills and their consequences."[10] Do you support such an international agreement? Should the United States be a party to such an agreement? Why or why not?

**Question 16-9:**   Refer back to Figure 16-2. Did this disaster affect the world price of gold? How are the costs of disasters like this reflected in the price of gold? Should they be? Why or why not?

## Gold Mining with Mercury: The Amalgamation Method

*Amalgamation* is the alloying and collection of gold from ore using droplets or coatings of mercury. The process is a simple one. Crushed gold ore is passed over surfaces coated with

---
[10]  Ibid.

mercury, to which the gold binds. The mercury–precious metal amalgam is then collected and heated to vaporization to separate the precious metals. Ideally, the mercury would be condensed in a retort for reuse, but considerable loss of mercury (and precious metals) is commonplace during the many steps involved in processing the ore. Accordingly, vaporized mercury during amalgamation is a significant source of regional mercury pollution.

In the United States, the average concentration of gold in ores is around 6 grams per tonne of rock! Large amounts of mercury were used in gold mining. In the Carson River Valley of Nevada, much of the sediment is presently mercury-contaminated due to historic gold mining.

In the developing world, and especially in Brazil, French Guyana and in the former Soviet Union, gold mining and processing using mercury is generating significant quantities of mercury pollution. Studies report fish in the Amazon River to be contaminated with mercury. In Brazil, labor is cheap, and the amalgamation method is simple: crush the ore, amalgamate it, and then burn the amalgam to vaporize the mercury and release the gold. The potential for exposure is high because mercury has a high *vapor pressure,* which means that whenever ore-concentrates containing mercury are heated, the mercury is vaporized.

Ironically, miners have low exposure to mercury since the mercury emitted is mainly less-toxic elemental mercury. However, elemental mercury may be converted to methylmercury, an extremely toxic chemical in soils and waters, and may thereby enter the food chain.

As a result, residents who eat fish from the Amazon are exposed to a much higher risk from mercury poisoning than miners, since methylmercury is far more toxic than elemental mercury. Mercury is widespread in the global environment, and most of it is from human activity like gold mining and coal burning.

**Question 16-10:** As you have seen, except for the electronics industry, most gold purchases are discretionary. To what extent should consumers consider the environmental record of gold producers before buying their product? Would you alter a purchase because of an adverse environmental and/or social impact of the activity? Give logical reasons for your response.

## GOLD RECYCLING

**Question 16-11:** "Gold is the single worst environmental purchase most consumers will make in their lives."[11] To what extent can recycling of gold reduce the need to develop new mines, which as you have seen often result in serious environmental pollution?

---

[11] www.retroworks.com/good_ideas.htm.

**Question 16-12:**   Review Figure 16-1. How much gold in total has been produced?

According to industry reports, almost 85 percent of all gold ever produced has been accounted for.

Gold demand for jewelry is increasing, especially in India with the growth of that country's middle class. Jewelers bought 80 percent of gold in 2005, although jewelry demand fell slightly (about 3%) in 2006 due to the sharp rise in gold price. Total gold demand in 2005 was approximately 4,300 tons.

**Question 16-13:**   Based on 2005 demand figures and the total gold produced and accounted for, could already-existing gold stocks satisfy demand by recycling alone? What additional information, if any, would you like to have to answer this question?

Not all gold can be easily recovered by recycling. As you have seen, it is difficult to recover gold from electronic products that are proliferating globally because of the extremely small amount of gold in each piece. One drawback to recovery of gold from electronics is the cyanide solution used to plate the gold (see above). New electroplating techniques using non-toxic organic chemicals instead of cyanide promise to make electronics recycling more feasible.

**Question 16-14:**   There is a bigger question here: Do people anywhere have the right, on an interconnected Earth, to carry out activities that degrade large parts of the planet on which many ensuing generations will have to live? Give logical reasons for your answer.

**Question 16-15:**   Is gold mining consistent with the goals of sustainable societies? Why or why not?

**Question 16-16:** Summarize in a paragraph or two the main points of this Issue.

## FOR FURTHER THOUGHT

**Question 16-17:** Research the impact of amalgamation gold processing on the Amazon region of Brazil. How is it affecting indigenous people?

**Question 16-18:** How does California handle pollution from 19th century gold mining?

# Issue 17

## PERSISTENT ORGANIC POLLUTANTS (POPS)

### KEY QUESTIONS

- What are POPs?
- What is the "Dirty Dozen Plus Nine?"
- What international agreements address the problems posed by POPs?
- What quantities of POPs are in the environment presently, and how will this change in the future?
- What are PCBs, how were they formed, and what threats do they pose to the global environment?

### THE CONVENTION ON PERSISTENT ORGANIC POLLUTANTS (POPS)

On May 23, 2001, U.S. officials signed the Convention on Persistent Organic Pollutants (POPs) in Stockholm, Sweden. Under the Convention, countries agree to reduce and eventually cease the production, use, and release of the twelve POPs of greatest hazard to the global environment. The agreement further sets up a scientific review process whereby additional chemicals may be added to the treaty as warranted.

The Stockholm Convention targets a group of POPs, informally called the "dirty dozen," shown in Table 17-1.

The POPs agreement begins global action to reduce and eventually eliminate these chemicals. The Convention took effect in May 2004 after ratification by fifty nations. By 2012, virtually every nation on Earth had ratified the Convention except Iraq, Saudi Arabia, Afghanistan, Italy, Malaysia, Western Sahara, Turkmenistan, Uzbekistan, and the United States of America.

In 2009, the Convention added nine more POPs to its list. They are:

| Chemical | Source |
|---|---|
| Alpha-and-beta Hexachlorocyclohexane | Pesticide |
| Chlordecone | Pesticide |
| Hexabromobiphenyl and hexabromobiphenyl ether | Pesticide |
| Lindane | Industrial chemical |
| Pentachlorobenzene | Pesticide and industrial chemical |
| Perfluorooctane sulfonic acid and its salts | Industrial chemical |
| Endosulfan and isomers of same | Pesticide |
| Tetrabromodiphenyl ether | Industrial chemical |
| Pentabromodiphenyl ether | |

**TABLE 17-1** ■ The Dirty Dozen and Their Origins (Sources: EPA and World Bank).

| 1 = Pesticide     2 = Industrial Chemical | 3 = Combustion Byproduct |
|---|---|
| Chemical | Comments |
| aldrin[1] | Fatal dose, 5g, adult male: fatal dose for women and children less. |
| hexachlorobenzene[1,2,3] | Can be lethal: has been found in food of all types. |
| chlordane[1] | Toxic to many animals. Human exposure is mainly by air. |
| mirex[1] | Used against fire ants. Very stable and persistent. Possible human carcinogen. |
| DDT[1] | Has been detected in breast milk. May harm infants. |
| toxaphene[1] | Most widely used pesticide in the United States in 1975. Possible human carcinogen. |
| dieldrin[1] | Mutagenic. Second most common pesticide detected in a U.S. survey of pasteurized milk. |
| polychlorinated biphenyls (PCBs)[2,3] | Suppress human immune system; probable human carcinogen; readily transferred in breast milk. |
| endrin[1] | Toxic, but can be metabolized, so little bioaccumulation. |
| polychlorinated dibenzo-p-dioxins (dioxins)[3] | Seven types out of 75 are mutagenic, carcinogenic. |
| heptachlor[1] | High doses fatal to birds, mammals; low doses mutagenic. |
| polychlorinated dibenzo-p-furan (furans)[3] | 135 different types; possible human carcinogen; can accumulate in breast milk. |

**Question 17-1:** Research one of these chemicals and report on its uses and toxicity.

# WHY ARE POPS OF GLOBAL CONCERN?

The World Bank reports that, of all the pollutants released into the environment by human activity, POPs are among the most dangerous. POPs are of global concern because there is firm evidence of *global* transport of these substances, by air and water, to regions where *they have never been used or produced* such as the North American Arctic. The ensuing threats posed to the entire global environment prompted the POPs agreement. The main threats POPs pose center on their tendency to (1) persist in the environment, (2) bioaccumulate in the food chain, and (3) adversely affect human and animal populations. People are mainly exposed to POPs by eating contaminated foods. In humans and other mammals, POPs can be transferred through the placenta and breast milk to developing offspring. We discuss this phenomenon with *orca*, killer whales, below.

POPs are extremely toxic. They cause a range of harmful effects among humans and animals, including cancer, birth defects, damage to the nervous system, reproductive disorders, disruption of the immune system, and even death. POPs can damage the reproductive and immune systems of exposed individuals as well as their offspring. Some POPs are *endocrine disrupters*. An endocrine disruptor is a chemical that interferes with the function of the endocrine system. It can mimic a hormone, block the effects of a hormone, or stimulate hormone production. Table 17-2 shows persistence in agricultural soils of a number of POPs.

**TABLE 17-2** ■ Persistence of POPs in Agricultural Soils (Courtesy of Ralph G. Nash and Edwin A. Woolson/AAAS.)

| Chemical | Years since Treatment | % Remaining |
|---|---|---|
| Aldrin | 14 | 40 |
| Chlordane | 14 | 40 |
| Endrin | 14 | 41 |
| Heptachlor | 14 | 16 |
| Toxaphene | 14 | 45 |
| Dieldrin | 15 | 31 |
| DDT | 17 | 39 |

## COSTS OF EXPOSURE TO POPS

In June 2000, the National Academy of Sciences, the federal government's premier science advisory organization, issued a report titled "Scientific Frontiers in Developmental Toxicology."[1] Here is a portion of the Executive Summary:

> "Of approximately 4 million births per year in the U.S., *major* developmental defects [occur] in approximately 120,000 *live-born* (our emphasis) infants."

**Question 17-2:** According to the NAS, what percent of live births results in major developmental defects?

> "At present, the causes of the majority of developmental defects are not understood; however, it *is* known that prenatal exposure to certain chemicals (such as POPs) and physical agents (such as radiation) found in the environment can cause developmental defects. Approximately 3% of all developmental defects are . . . caused by exposure to toxic chemicals and physical agents, including environmental agents, and almost 25% of all developmental defects might be due to a combination of genetic and environmental factors."

It is no longer controversial that environmental toxins impose a significant cost upon humans, especially fetuses and infants. If we were to assume a minimal cost of each human life at $1 million (based on awards in liability lawsuits), then the *minimal costs* of such toxins in the environment could exceed $3.6 billion annually in the United States alone, and is likely to be much more."

## POPS, ORGANOCHLORINES, AND THE PRECAUTIONARY PRINCIPLE

Recall the precautionary principle from the first section of this book. Restated, it reads, "Human society should avoid practices that have the potential to cause severe damage, even absent absolute scientific proof of harm." There is no better example of the need for application of the precautionary principle than that group of POPs produced when chlorine gas reacts with organic compounds. The compounds produced are called *organochlorines*.

---

[1] National Research Council: Committee on Developmental Toxicology, Board on Environmental Studies and Toxicology, Commission on Life Sciences. 2000. *Scientific Frontiers in Developmental Toxicology and Risk Assessment* (Washington, DC: National Academy Press.

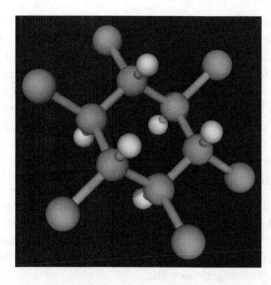

**FIGURE 17-1** Molecular structure of an organochlorine. (Courtesy of Laguna Design/Science Photo Library/Corbis Images.)

Chlorine gas is greenish, heavy, extremely reactive, and does not occur in nature, not even in that cauldron of creation, active volcanoes. Chlorine gas readily reacts with any organic chemical it encounters, producing bleaches, disinfectants, insecticides and pesticides, and, incidentally, hundreds of additional by-products.

Organochlorines may be chemically organized in two fashions. One type, the *aromatics,* contains structures called benzene *rings* (see Figure 17-1). The other type, *aliphatics,* from the Greek for *fat* (fats have the chain structure as well), consists of *chains* of carbon atoms.

To give you an idea how these chemicals are named, consider PCBs. The chemical name for this family of compounds is *polychlorinated biphenyls.* A biphenyl is made of two linked benzene rings: If chlorines are added, it becomes a polychlorinated biphenyl. Some other extremely toxic chemicals are similar in structure to PCBs. DDT (*di*chloro-*di*phenyl*tri*chloroethane), for example, has one chlorine attached to each of two benzene rings, in turn, attached to a trichloroethane.

The aliphatics, since they closely mimic the structure of fats, are highly bioaccumulative (that is, they build up in fatty tissue). For example, hexachlorobutadiene has a bioaccumulation factor of up to 17,000. Others have ratios approaching 70,000.

**Question 17-3:** How many chlorines are found in hexachlorobutadiene?

## HOW DOES CHLORINATION AFFECT ORGANIC CHEMICALS?

Chlorination radically changes the properties of organic compounds. It can increase the stability of the compound so that it may persist in the environment for decades or centuries. This property has been of great value to industry over the past hundred years and provided us with our first "safe" refrigerant, freon. The first generation of refrigerators used ammonia as a refrigerant, and ammonia is toxic when ingested. Freons were believed to be inert.

Since chlorinated hydrocarbons, by and large, are not naturally produced, there are few mechanisms that remove or degrade them once formed.

The addition of chlorine gas to organic compounds also increases their reactivity. And perhaps most important, adding chlorine to organic chemicals *increases their solubility in fats and oils.* This means they can bioaccumulate in fatty tissues of animals and can be passed on from generation to generation by mother's milk. They can also become concentrated in larger animals by the simple process of eating smaller ones, a process called *biomagnification.* Thus, humans can concentrate organochlorines in their bodies by eating contaminated fish, and seabirds and marine mammals can, and do, experience the same effect.

## How Safe Are Organochlorines?

Only a few hundred of the thousands of organochlorines produced by industry have been tested by scientists, and virtually all of them have been found to damage one or more of the following processes: fetal development, brain function, the immune system, the endocrine system, and/or sperm production and development. They may be *mutagenic* (causes genetic mutations) and *carcinogenic* as well.

Organochlorines cause changes in bone composition including reduced bone mineral density. Organochlorines have also been implicated in other bone diseases including periodontitis, a widespread disorder of the gums and bones around the teeth.[2]

**Question 17-4:** Should women with a tendency for bone-density loss or people with periodontitis be encouraged to be tested for organochlorines? Why or why not? Do you think this is routinely done?

Finally all of these harmful effects can occur at *parts-per-trillion concentrations*, which is described as "equivalent to one drop in a train of railroad tank cars 10 miles long."[3]

**Question 17-5:** Reread Jefferson's quotation on corporations, from "Basic Concepts and Tools," (p. 15). Do you believe that corporations should be held liable for any damage that their actions cause, in violation of the precautionary principle? How? Cite evidence to support your view, recalling that opinions uninformed by evidence are of little value in scientific inquiry.

## PCBs

Recall the "dirty dozen" from Table 17-1. One of these was polychlorinated biphenyls (PCBs). PCBs are a group comprising over 200 structures. Their formula is complex, and their atomic weight depends on the number of chlorine atoms in the structure. PCBs do not exist naturally on Earth. They were first synthesized during the late nineteenth century.

---

[2] National Institute of Health (NIH), www.pubmedcentral.nih.gov/articlerender.fcgi?artid=1331997.
[3] Thornton, J., in *Pandora's Poison: Chlorine, Health and a New Environmental Strategy.* 2000. (Columbia University, New York, New York.

Because of their stability when heated, they were widely used in electrical capacitors and transformers. In the 1960s, scientists began to report toxic effects on organisms exposed to PCBs, and by 1977, the manufacture of PCBs was banned in the United States, the United Kingdom, and elsewhere. By 1992, 1.2 million metric tonnes (2.6 billion lbs) of PCBs were believed to exist worldwide. As much as 370,000 metric tonnes (810 million lbs) could have been dispersed into the environment.

PCBs can be destroyed by incineration, but the process is expensive. Around 15 percent of PCBs in soils reside in developing countries, mostly from shipments of waste contaminated with PCBs from developed countries.

**Question 17-6:** What role, if any, should wealthy countries play in neutralizing PCBs in developing countries? Justify your answer by listing and defending your reasons.

PCBs are relatively heavy molecules (average atomic weight of around 360 g/mole) and are relatively insoluble in water. While concentrations in seawater can reach 1 part per million (ppm), PCBs typically concentrate in sediments. From there, they enter the food chain through the activities of organisms called sediment, or deposit, feeders. These creatures eat sediment, extract organic matter, and excrete the rest. Like other organochlorines, PCBs are *lipophilic* and thus tend to accumulate in the fatty tissues of animals. If other animals eat the deposit feeders, the PCBs are not metabolized and become more concentrated in the animal's fat (biomagnified). Concentrations exceeding 800 ppm have been measured in the tissues of marine mammals. According to the Environmental Research Foundation, this would qualify the creature for hazardous waste status!

PCBs are widespread pollutants and have contaminated most terrestrial and marine food chains. They are extremely resistant to breakdown and are known to be carcinogenic. PCBs have been linked to mass mortalities of striped dolphins in the Mediterranean, to declines in *orca* (killer whale) populations in Puget Sound, and to declines of seal populations in the Baltic.

While PCBs may threaten the entire ocean, the northwest Atlantic is believed to be the largest PCB reservoir in the world because of the amount of PCBs produced in countries that border the north Atlantic. PCBs have been shown to cause liver cancer and harmful genetic mutations in animals. PCBs may inhibit cell division, and they have been implicated in reduction of plant growth and even mortality of plants. According to a report edited by Paul Johnston and Isabel McCrea for Greenpeace UK,

> Since the rate at which organochlorines break down to harmless substances [has been] far outstripped by their rate of production, the load on the environment is growing each year. Organochlorines (including PCBs) are arguably the most damaging group of chemicals to which natural systems can be exposed.[4]

## Bioaccumulation and Biomagnification of PCBs

POPs accumulate in the body fat of living organisms, as you have seen, and become more concentrated as they move from one creature to another. When contaminants found in small amounts at the bottom of the food chain biomagnify, they can pose a hazard to predators at the top of the food chain.

---

[4] Johnston P., & I. McCrea. 1992. *The Effects of Organochlorines on Aquatic Ecosystem.* London: Greenpeace International.

A 1997 study found that caribou in Canada's Northwest Territories had up to ten times the levels of PCBs as the plant matter on which they fed. PCB levels in the wolves that ate the caribou were nearly six times higher still.

J. Cummins, in a 1988 paper in *The Ecologist,* concluded that adding 15 percent of the remaining stock of PCBs to the ocean would result in the extinction of marine mammals.[5]

## Highly Exposed Populations[6]

The long-distance transport of POPs toward the poles has contaminated the Arctic food web. Indigenous peoples of the Arctic experience a high intake of organochlorines from consuming a traditional diet featuring marine mammals, which have accumulated high levels of organochlorines from their food. Populations who have a diet rich in fish from contaminated waters, such as residents of the Great Lakes region (in the United States and Canada) and of the shores of the Baltic Sea, have a high intake of organochlorines. Children can have a higher intake of organochlorines than adults because of their comparatively high food intake. In addition, exposure of the developing young is of great concern because these stages of life are most vulnerable to the toxic effects of POPs. Nursing infants have a particularly high intake of organochlorines because of bioaccumulation in breast milk.

**Question 17-7:**   Should pregnant women, or those desiring to become pregnant, be encouraged to be tested for organochlorine contamination? Why or why not?

## Synergistic Effects: PCBs and Mercury

Combining exposure to toxic chemicals can multiply the harmful effects of each: This is called *synergism.* One particularly toxic mercury compound, methylmercury, biomagnifies powerfully as it goes up the food chain. Figure 17-2 shows this effect from the Florida Everglades. Water concentration of 0.1 part per trillion (ppt) was biomagnified to 2,000 ppt in plants, and so on. Mercury exhibits synergistic effects with PCBs and other POPs. A study of children born to mothers who consumed fish from Lake Ontario showed that prenatal PCB and mercury exposures interacted to reduce performance of three-year-old children on certain tests.

## PCBs and Orcas in Puget Sound

Even though soldiers during World War II used them for target practice, orca have become a symbol of the Pacific Northwest. In 1999, Dr. P. S. Ross, a research scientist with British Columbia's Institute of Ocean Sciences, took blubber samples from forty-seven live killer whales and found PCB concentrations of 46–250 ppm, up to 500 times greater than those found in humans. Ross concluded, "The levels are high enough to represent a tangible risk to these animals."[7]

Ross compared the orca population he studied with the endangered beluga whale population of the St. Lawrence estuary of eastern North America, in which a high incidence of

---

[5] J. E. Cummins, The PCB Threat to Marine Mammals. *The Ecologist,* Nov–Dec 1988.
[6] www.greenpeace.org.
[7] "Toxin Threatens a Wonder of the Northwest," by M. L. Lyke Special to the Washington Post, Monday, November 8, 1999, Page A9.

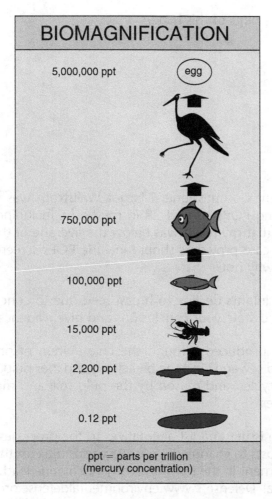

**FIGURE 17-2** Biomagnification of methyl–mercury in the Florida Everglades. (www.usgs.gov)

diseases have been linked to contaminants and which have shown evidence of reproductive impairment.

For the orca, the PCBs are likely passed from generation to generation. PCBs are highly fat-soluble and are concentrated in mother's milk. Ross said, "Calves are bathed in PCB-laden milk at a time when their organ systems are developing and they are at their most sensitive."[8]

While PCBs have been banned in the United States for nearly three decades, they are still being used in some developing countries. Accordingly, Ross speculates that PCBs in the Pacific could be derived from East Asian sources and could end up concentrated in the tissues of migratory salmon, which are a prime food source for the Orca. Ross's study was done in collaboration with the University of British Columbia, the Vancouver, British Columbia, Aquarium, and the Pacific Biological Station of British Columbia.

**Question 17-8:** Discuss the topic of POPs from the perspective of sustainability and sustainable societies.

---

[8] "Killer Whales are Full of Toxic Chemicals, New Study Says PCBs Make Popular Orca Prey to Menacing Diseases." *Seattle Post-Intelligencer,* 10/25/99.

**Question 17-9:**   Summarize the main points of this Issue.

## FOR FURTHER THOUGHT

**Question 17-10:**   European Environment Commissioner Margot Wallstrom was screened in 2003 for 77 toxins including POPs. "I had 28 in my body, including PCB and DDT," she said. "I was told that my result was below the average of the group tested."[9] Do you think health-care providers should include POPs screening for the entire population? Why or why not?

**Question 17-11:**   Should women with infants decline to breast-feed due to concerns about POPs? Research this issue and state your conclusions and give evidence.

**Question 17-12:**   Organochlorines are produced through the chlorination of drinking water, the purification of treated sewage, and the bleaching of paper pulp. Chlorinated pulp liquids cannot be recycled and reused by the pulp mill and must be discharged into local water bodies.

Research any of these issues and list alternatives to the processes used that rely on chlorine. Assess the costs to change the process versus the cost that is already being borne by the environment in the form of its organochlorine load. Consult the web-sites of Environmental Defense, www.environmentaldefense.org; the EPA, www. epa.gov; and those of major chemical producers like duPont for data on POPs. Also see *Pandora's Poison*, by J. Thornton (MIT Press, Cambridge, MA, 2000).

**Question 17-13:**   Some critics assert that Greenpeace scientists (whom we have cited as references) may be too biased to provide reliable data on PCBs. Research PCBs on the Internet, in chemistry texts, or in other sources. Cite any differences between the treatment of these chemicals by Greenpeace and the sources you found.

**Question 17-14:**   Here are some facts about the distribution of organochlorines:

| | |
|---|---|
| Organochlorine compounds found in the tissues and fluids of the general North American population | 193 |
| Organochlorine contaminants so far discovered in the Great Lakes | 83 |
| Organochlorine by-products produced by the chlorination of drinking water and waste water | 40 |
| Organochlorine by-products of hazardous waste incinerators | 31 |

Research these issues and conclude to what extent the hazards posed by organo-chlorines are offset by the value provided to society by the processes listed.

**Question 17-15:**   Under the Convention (www.pops.int) may some POPs continue to be used? Why?

---

9   www.cancerpage.com/news/article.asp?id=7077.

# Issue 18

## GLOBAL GRAIN PRODUCTION: *CAN WE BEEF IT UP?*

### KEY QUESTIONS

- How did humans domesticate wild grains?
- What are global trends in grain production?
- What are the principal uses of grain in the United States and the world?
- Can global food supplies feed a growing population?

### WHY WE EAT

Reduced to its basic chemical components, the human body is carbon, hydrogen, oxygen, nitrogen, and a few other elements. These combine to form more complex, highly organized structures called molecules, such as proteins, lipids, carbohydrates, and nucleic acids. These molecules interact to build even larger and more complex structures: cells, tissues, organs, organ systems, and, ultimately, the unique individual.

Such a system of highly organized, integrated parts, however, does not exist without cost. It is subject to the same laws of physics that affect any matter. Among these is the *second law of thermodynamics,* which states that disorder (entropy) in the universe is increasing and that whenever energy is converted from one form to another, some of the input energy is lost in the process of conversion, resulting in less output energy. Practically speaking, in the natural state, complex, organized systems tend to become simple and unorganized unless energy (in our case, food) is added to the system to keep it organized and complex. Inadequate nutrition leads to undernourishment or malnutrition. Globally, perhaps 1 billion people are malnourished.

Table 18-1 compares foods and feed grains for China, the U.S., and the world.

**Question 18-1:** Contrast the American diet with the typical Chinese diet.

TABLE 18-1 ■ Selected Foods Available to U.S. and Chinese Consumers
(kg per capita per year.)

| Food | U.S. | China | World |
|---|---|---|---|
| Food Grain | 92 | 387 | 158 |
| Vegetables | 191 | 198 | 167 |
| Fruits | 135 | 35 | 60 |
| Fiber | 12 | 33 | —* |
| Meat, Fish | 91 | 62 | 56 |
| Dairy | 272 | 7 | 79 |
| Fats, Oils | 31 | 5 | 13 |
| Sugars | 72 | 7 | 25 |
| Nuts | 9 | — | 1 |
| Animal Protein | 11 | .8 | —** |
| Total Food | 908 | 715 | 567 |
| Animal Feed | 816 | 70 | 150 |

**Percent of total calories; *grams per day. Sources: Pimentel, David, and Marsha Pimentel, 2003, *Am J Clin Nutr* 2003;78(Suppl):660S–3S, and Campbell, C., and T. Campbell, *The China Study*. Benbella Books, Dallas, TX 2006.

# WHAT WE EAT: GRAINS AND THE ORIGIN OF AGRICULTURE

Even though humans are omnivorous and eat a variety of foods, we are extremely dependent on only a few crops: corn (maize), wheat, rice, barley, and soy. Why do we depend so heavily on these relatively few plant species? For one thing, these plants have significant protein and carbohydrates. For another, most wild plants that we could possibly domesticate are useless to us as food—they may be indigestible, very low in food value, or hard to prepare. In fact, most wild biomass is wood or leaves, which we can't digest.

Plant domestication has been defined as "growing a plant, and thereby consciously or unconsciously causing it to change genetically from its wild ancestor in ways that make it more useful to humans."[1] The history of plant domestication is a fascinating one. Here's how domestication may have come about.[2]

Studies of plant remains at human habitation sites tell us when humans first domesticated plants, since domesticated grains are much different from their wild ancestors. Grain domestication was preceded by a major climate warming, resulting from the last glacial withdrawal around 11,000 years ago. This warming led to an expansion of the habitat for the wild grains that were ultimately to be domesticated, making it easier for humans to collect them.

According to carbon-14 dating, grain production probably arose first in the Fertile Crescent of the Near East, where the domestication of wheat from wild varieties occurred by 8500 BCE. It quickly spread along the great east-west axis of Eurasia. The domestication of rice began in China sometime around 7500 BCE. In the New World, corn (maize) was first domesticated by 3500 BCE.

When humans began to grow crops from wild varieties, they would naturally select the varieties that would best suit their needs. Recall that there is a wide variation in appearance of wild fruits like blueberries, blackberries, and strawberries. Generally, early farmers probably selected according to size—the bigger the fruit, the bigger the offspring, or so they may have thought. Cultivated peas, for example, are ten times bigger than their wild

---

[1] Diamond, Jared. 1997. *Guns, Germs and Steel* (New York: W.W. Norton), p. 114.
[2] Ibid.

ancestors, which hunter-gatherers must have been collecting for centuries. To take another example, the oldest corncob from MesoAmerica was 1 to 2 centimeters long. By 1500, farmers were growing 15-centimeter varieties, and modern cobs may exceed 30 centimeters. Domestication of the plants we take for granted was a process that likely developed over centuries. In fact, it is still occurring. Strawberries were not domesticated until the Middle Ages, and pecans were not domesticated until the middle of the nineteenth century.

Some of the ancestors of the plants we cultivate—eggplant, potatoes, and almonds, for example—were poisonous. These plants do have a *recessive* gene, however, that will not grow poisonous offspring. In the wild, such a gene is likely to stay recessive since animals will likely gobble up any edible varieties. But humans, after a bit of admittedly risky trial and error, would eventually select plants with the recessive gene and begin to grow only forms that displayed the nonpoisonous characteristics they needed. Recessive genes have been of great value to human society. Wild wheat plants have a dominant gene that causes the stalk to disintegrate when the plant's seeds are ripe. This allows the seeds to be spilled upon the ground, permitting them to germinate. But a recessive variety exists that lacks this gene. These stalks don't disintegrate, and this is the wheat plant that early farmers selected.

In short, here is why we became dependent on so few grains:

- Their ancestors were edible and plentiful in the wild.
- They grew quickly and were easily collected and stored.
- They contained protein and were rich in carbohydrates.

Thus, as a result of millennia of trial and error, by the days of the Roman Empire almost all modern grains were being cultivated in some form somewhere.

## THINKING CRITICALLY ABOUT GLOBAL GRAIN PRODUCTION

Over most of human history, farmers have increased agricultural output mainly by plowing up forests and natural grasslands. Limits of geographic expansion were reached long ago in densely populated parts of India, China, Java, Egypt, and Western Europe. *Intensification* of production—obtaining more output from a given area of agricultural land—has thus become a "growing" necessity. Particularly in Asia, this has been achieved through producing several crops each year in irrigated "agro-ecosystems" using new, fast-growing crop varieties—the so-called "Green Revolution."

However, growth rates in grain yields have slowed in both developed and developing countries. And future increases in food production may become more difficult because environmental and social factors must now be taken into account as we develop new crop technologies. An illustration of the latter is the growing controversy over genetically modified foods, which we briefly discuss in Issue 28.

Let's start our analysis by considering the following quotation from Bailey, 1995.

Food is more abundant and cheaper today than at any other time before in history. Per capita grain supplies have increased 24 percent since 1950, while food prices have plummeted by 57 percent since 1980.

Food production has outpaced population growth since the 1960s. The increase in food production in poor countries has been more than double the population growth rate in recent years.[3]

Now, consider this quotation from Lester Brown of Earth Policy Institute.

The 2006 world grain harvest is projected to fall short of consumption by 61 million metric tons (mmt), marking the sixth time in the last seven years that production has failed to satisfy demand. As a result of these shortfalls, world carryover stocks at the end of this crop year are

---

[3] Bailey, R. (ed.) 1995. *True State of the Planet* (New York: Free Press), p. 50.

projected to drop to 57 days of consumption [or 319 mmt], the shortest buffer since the 56-day (low) in 1972 that triggered a doubling of grain prices.

World carryover stocks of grain, the amount in the bin when the next harvest begins, are the most basic measure of food security. Whenever stocks drop below 60 days of consumption, prices begin to rise.[4]

Janet Larsen of the Earth Policy Institute had this to say about the 2011 harvest:[5] "The U.S. Department of Agriculture [reports] a global grain harvest of 2,295 million tons, up 53 million tons from the previous record in 2009. Consumption grew by 90 million tons over the same period to 2,280 million tons. Yet with global grain production actually falling short of consumption in 7 of the past 12 years, stocks remain worryingly low, leaving the world vulnerable to food price shocks."

Brown further notes, ". . . world grain demand, traditionally driven by population growth and rising incomes, is also now being driven by the fast-growing demand for grain-based ethanol for cars."[6]

**Question 18-2:**  Are the Bailey and Brown statements consistent with each other? What additional information would you need in order to choose which statement(s) is/are accurate?

Figure 18-1 shows world grain production from 1960–2011.

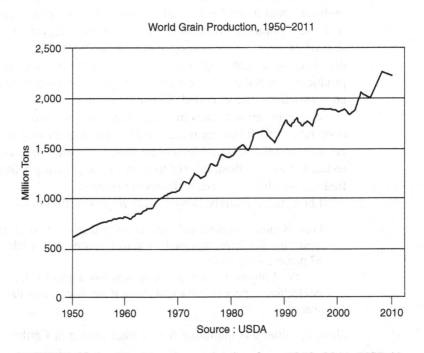

**FIGURE 18-1**  World grain production from 1950–2011 (USDA).

---

[4] Earth Policy Institute, www.earth-policy.org/indicators/C54/grain_harvest_2006.
[5] Earth Policy Institute, http://www.earth-policy.org/indicators/C54/grain_2012.
[6] Earth Policy Inst, op cit.

**Question 18-3:** Describe the change in global grain production between 1960 and 2011.

**Question 18-4:** In 1960 world population was 3.03 billion. In 2011, world population was 6.9 billion. What was per capita grain production in 1960 and 2011?

**Question 18-5:** Research the production of ethanol. How do you think ethanol production for a transport fuel will affect grain supplies? What evidence would you need to fully answer this question?

We briefly consider the impacts of producing ethanol from corn in "For Further Thought" at the end of this Issue.

## GRAINLAND AREA

In the United States, many rural and suburban dwellers are used to seeing farms being converted into residential subdivisions with miles and miles of tract homes serviced by highways and power lines. (The "Poster Child" for this effect may well be California's Central Valley.) At the same time, there has been intensification of agricultural land use around some major foreign cities (and even within cities), particularly for high-value perishables such as dairy and vegetables but also to meet subsistence needs. Globally, grain area harvested per capita has fallen from 0.21 ha in 1960 to 0.1 ha in 2010.[7]

Figure 18-2 shows global grain price changes since 2005.

**Question 18-6:** Refer to Figure 18-2 and the quotations from Bailey and Brown on page 177–178. Which one does your assessment tend to support? Explain your reasoning.

Although the per capita grainland harvested has decreased by 50 percent from 1960 to 2012, this decrease has been offset by compensatory increases in output per hectare achieved by high-yield farming. Called the Green Revolution, the development and use of new strains of wheat and rice along with greater irrigation (Figure 18-3), input of fertilizers, pesticides, herbicides, fungicides, and so forth, drastically increased crop yields.

---

[7] www.worldwatch.org/node/554.

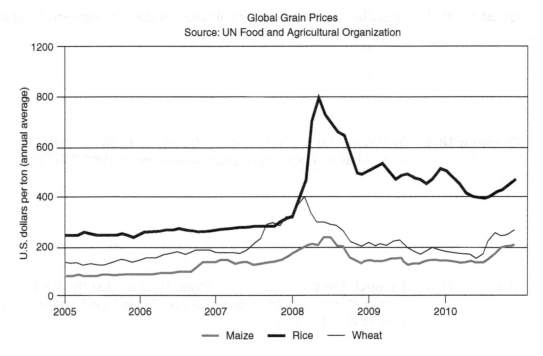

**FIGURE 18-2**   Changes in global grain prices since 2005 (UN FAO).

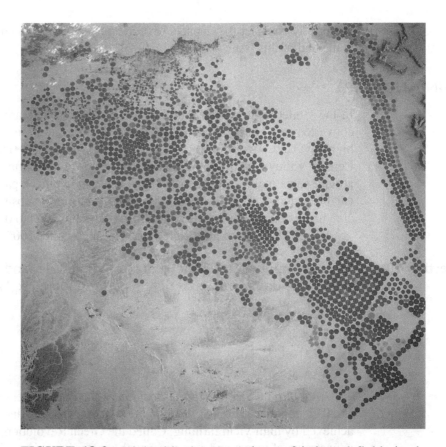

**FIGURE 18-3**   High altitude space photo of irrigated fields in the Arabian desert. Irrigation water is from 1,300 m deep aquifers. Circles are approximately 0.5 miles (550 m) in diameter (Courtesy of Corbis Images.).

## GRAIN AND MEAT PRODUCTION

World meat production was 296 million tonnes (mt) in 2011. It was 44 mt in 1950.

**Question 18-7:** Calculate the per-capita meat production for 1950 and for 2011, then characterize the change in global per capita meat production between 1950 and 2011. Global population was 2.56 billion in 1950, and 6.98 billion in 2011.[8]

Economic growth in Asia has helped fuel this increase. Annual per capita meat consumption ranges from about 124 kilograms in the United States to 75 kg for Greece, 53.4 kg in China to around 3.2 kilograms in India. Some 15 billion livestock (11 billion of which are poultry) exist at any one time to satisfy this demand[9] (see Figure 18-4).

**Question 18-8:** Does an increase in meat consumption signal improving global nutrition? Explain your answer.

Much of the world's grain harvest is fed to cattle and other livestock. In 1960, 294 out of 822 million tons of grain were used as animal feed. In 2006, the Earth Policy Institute estimated that 714 mt (million tons) out of a total global grain harvest of 1,984 mt were fed

**FIGURE 18-4** Cattle feedlots near Lubbock, TX. (Courtesy of Richard Hamilton Smith/Corbis Images.)

---

[8] U.S. Dept. of Agriculture, www.fas.usda.gov/psdonline/circulars/livestock_poultry.pdf.
[9] For details see http://chartsbin.com/view/bhy.

to animals. China News reports that as of 2011 70% of Chinese corn (maize) production was used as feed.[10]

**Question 18-9:** About what percentage of the global grain harvest was fed to animals in 2006?

According to the U.S. Department of Agriculture, nearly 25% of corn grown in 2011 was used to produce ethanol. High-protein "corn cakes" a byproduct of ethanol production, is now a major agricultural export. Additionally, about 60% of all grains produced during the late 2000s was used as animal feed.

Pimentel and Pimentel estimate that a switch to grass-fed livestock in the United States would allow about 130 million tons of grain to be diverted to human consumption, which could feed about 400 million people annually.[11]

**Question 18-10:** Should we discourage meat consumption or ethanol production, or encourage grass feed instead of grain for livestock to free grain to feed the world's hungry? Explain your answer.

**Question 18-11:** Prime farmland is disappearing around large U.S. cities. The land is being used for housing, roads, office buildings, and shopping centers. Some have questioned whether it is "economical" to convert farmland around cities into enormous parking lots for grocery stores in which food grown thousands of miles away is sold. Do you concur? Discuss and defend your answer using critical thinking principles.

**Question 18-12:** Under what conditions will it be possible to feed a global human population of 9+ billion? Speculate on what the planet's environment would be like in such an eventuality.

**Question 18-13:** Discuss Issue 18 from the perspective of sustainability and sustainable societies.

---

[10] www.xinhuanet.com/english/.
[11] www.news.cornell.edu/releases/Aug01/corn-basedethanol.hrs.html and Pimentel, D., & M. Pimentel. 1996. *Food, Energy, and Society* (Niwat, CO: University Press of Colorado).

**Question 18-14:** Summarize the main points of this Issue.

## FOR FURTHER THOUGHT

**Question 18-15:** Assess the impact on global grain supplies of the growth in Chinese pork consumption.

**Question 18-16:** The following is from the website of the Natural Resources Defense Council (http://www.nrdc.org/food/wasted-food.asp):

Food is simply too good to waste. Even the most sustainably farmed food does us no good if the food is never eaten. Getting food to our tables eats up 10 percent of the total U.S. energy budget, uses 50 percent of U.S. land, and swallows 80 percent of freshwater consumed in the United States. Yet, 40 percent of food in the United States today goes uneaten. That is more than 20 pounds of food per person every month. Not only does this mean that Americans are throwing out the equivalent of $165 billion each year, but also 25 percent of all freshwater and huge amounts of unnecessary chemicals, energy, and land. Moreover, almost all of that uneaten food ends up rotting in landfills where it accounts for almost 25 percent of U.S. methane emissions."

How might food waste be reduced in the United States?

**Question 18-17:** According to the Organic Trade Association, "organic food is healthier and safer for consumers than nonorganic products." Demand for organically grown food is increasing in the United States and the European Union at 15 to 20 percent per year. Although organic food usually costs more than conventional food in supermarkets, economics of scale are helping to lower the price differentials. Consumers as of 2010 in the United States spent $27 billion on organic food. Research the issue at www.organicconsumers.org. Is organic food "better" than conventionally-produced food? How did you define "better" to answer this question?

**Question 18-18:** The use of corn to make ethanol motor fuel is increasing in the United States, driven by substantial government subsidies. However, according to Cornell University agricultural scientist Dr. David Pimentel, "Adding up the energy costs of corn production and its conversion to ethanol, 131,000 BTUs are needed to make 1 gallon of ethanol. One gallon of ethanol has an energy value of only 77,000 BTU. Put another way, about 70 percent more energy is required to produce ethanol than the energy that actually is in ethanol."[12] Much of the energy used to make ethanol is derived from gasoline, natural gas, and diesel fuel, and nearly half of the petroleum is imported. Research the issue, and comment on the long-term viability of an ethanol-based transport system derived from corn, based on Pimentel's statistics. Cite experts who disagree with Pimentel's assessment and state their views.

---

[12] Ibid.

# Issue 19

## SOILS AND SUSTAINABLE SOCIETIES

### KEY QUESTIONS

■ What are soils?
■ What is the status of the world's soils?
■ What can be done to ensure fertile soils in the future?
■ What is the relationship between soils and atmospheric $CO_2$?
■ How significant are soils to sustainable societies?

## THE NATURE OF SOIL

Soils, said Leonardo DaVinci, are the Earth's flesh. Even though they are critical to the survival of human society, we are just beginning to understand the role of soils in maintaining a healthy planet. Soils may have natural regulatory roles and benefits as yet unrecognized by humans. For example, recent discoveries show that soils contain a range of natural antibiotics produced by teeming soil microbes. The antibiotic streptomycin was itself discovered by soil investigators.

Soils are composed firstly of decomposed and disintegrated bits of rock and mineral matter. They vary greatly in age and rate of formation. While some soils may be millions of years old, most are far younger. Some pedologists (soil scientists) are expressing concern at the rapidity with which humans are degrading soils compared to the time it takes soils to form. Soil-making processes are notoriously slow. It can take from 200 to 1,000 years to form 2.5 centimeters (one inch) of topsoil. Some scientists wonder if soils will survive in their natural state long enough for us to truly understand them.

Soils also contain insoluble residues like iron- and aluminum-hydroxides that water can't dissolve. Most important, soils contain organic matter (humus) and uncountable mostly microscopic soil organisms. Many soil microbe communities remain unstudied in detail. In the words of one pedologist, soils are "living, breathing entities."[1] Their protection is critically important to agriculture, and perhaps of greater significance to society at large.

Humans degrade soils by

1. Paving over soils with asphalt and buildings (urbanization).
2. Accelerating soil erosion.
3. Poor agricultural practices.
4. Desertification.

---

[1] Schneiderman, J.S. 2000. *The Earth Around Us: Maintaining a Living Planet* (New York: W.H. Freeman), pp. 152, 153.

5. Irrigation practices leading to salt buildup
6. Adding toxins to kill soil microbes.

There are three general soil types, depending upon the climate under which the soil forms. They are *pedocals, pedalfers*, and *laterites*.

# URBANIZATION AND SOIL LOSS

Urbanization may ultimately be as great a threat to soils as poor agricultural practices. In testimony before Congress, American Farmland Trust (AFT) President Ralph Grossi said, "For the last 60 years America has sought to prevent the erosion of topsoil from wind and water, and through concerted action we have for the most part succeeded. Yet we continue to lose soil—to asphalt. In fact, the volume of topsoil lost to urbanization is roughly equivalent to the soil saved by the (federal government's) Conservation Reserve Program!"[2]

For example, the AFT estimates that by 2050 one-third of America's most important agricultural region, the Central Valley of California, will be paved over with suburbs. Professor Ronald Amundson of University of California-Berkeley wrote, "we are appalled when a 3,000-year-old tree is cut down, but our culture generally applauds the ingenuity of developers who can transform a 300,000-year-old soil into a golf course or housing tract."[3]

Development patterns help determine the extent to which urbanization causes soil loss. "Sprawl" development results in the loss of five to seven times as much sediment or soil compared to the forest it replaces and nearly twice the sediment loss of compact development or "smart growth" (see Issues 25 and 26). Forests permit little topsoil to erode, by contrast with conventional agriculture. In undeveloped forested land, (1) runoff is minimal, (2) plant cover and roots allow rainwater to seep slowly into soil, and (3) water filters slowly through ground to stream, which results in stable stream flow. Thus, erosion is minimal.

# ACCELERATING SOIL EROSION

Troubling many pedologists is the *rate* at which soils are disappearing. Pioneers reported rich pedocal soils 16 feet (5 m) deep in the prairies of Iowa, but only half of that remains. Floods in midwestern states over the past several decades have dumped untold millions of tonnes of fertile soils from farms into the Mississippi and other rivers, where they will eventually end up behind dams, on the river's flood plain, or in the Gulf of Mexico. Hurricane Floyd in 1999 dumped millions of tons of topsoil into the Atlantic from North Carolina farms and hog-processing operations, along with pesticides, fungicides, and fertilizer.

## Poor Agricultural Practices

According to the World Resources Institute, the two greatest sources of topsoil loss are sprawl development and improperly managed agricultural land.[4] The introduction of the steel plow to the U.S. Great Plains broke up the tough soil and removed the native long-rooted prairie grasses.

Cattle were introduced after the buffalo were slaughtered. These practices contributed to an episode of wind erosion unprecedented in the region's recent history: the so-called "dust bowl" years of the 1930s (Figure 19-1).

By 1990, poor agricultural practices had contributed to the degradation of 562 million hectares worldwide, about 38 percent of the roughly 1.5 billion hectares in cropland. Some

---

[2] Testimony in support of the Farmland Protection Program before the U.S. Senate Committee on Agriculture, Nutrition and Forestry. 7/21/99. www.ag.senate.gov.
[3] Ibid.
[4] www.igc.org.

**FIGURE 19-1** Loss of topsoil due to poor farming practices and wind erosion during the Dust Bowl era (Courtesy of Bettmann/Corbis Images.).

land was eroded severely enough to impair its productive capacity or to take it out of production completely. Since 1990, losses have continued, with as much as 5 million hectares lost to severe soil degradation annually.

**Question 19-1:** Put 5 million hectares into context (1,580 ha = 1 mi$^2$). How does this annual soil loss compare to the area of your state? (You can find state areas at www.enchantedlearning.com/usa/states/area.shtml.)

Soil erosion often results from water runoff caused by poor farming practices, and if erosion rates are not controlled, soil erosion can degrade water supplies.

## DESERTIFICATION AND SALT BUILDUP

Improper agricultural practices may cause long-term or irreversible land degradation. Intensive irrigation could turn one-third of Spain into desert. Similar practices have led to the degradation of vast tracts of the African Sahel, an area immediately south of the Sahara.

In arid regions, soluble elements such as selenium and sodium may accumulate in soils where there is insufficient rainfall to wash them below the water table. This process is called *salinization*. Salt buildup threatens to poison large areas of Canada's prairie provinces, for example.

Selenium is a nutrient in minute quantities but is toxic at high concentrations. Selenium-rich soils brought into agricultural production by large-scale irrigation can leach selenium into groundwater and rivers or cause selenium concentration in edible plants, where it can poison wildlife. In the Central Valley of California, for example, groundwater conditions are such that irrigation waters may pond near the land surface, eventually poisoning the soils with toxins like selenium. Indeed, the site of the original domestication of grains, the Fertile Crescent, in what is now Iraq, may have been permanently poisoned by irrigation thousands of years ago.

## ADDING TOXINS

Cultivated soils are different from uncultivated soils. Conventional strawberry and wine grape growers like to "sterilize" their soils to kill pests like nematodes, which can feed on the roots of young plants and thus reduce yields. One such chemical, until recently in more widespread use, is methyl bromide, an acute nerve toxin. It has been banned in many countries because of danger to agricultural workers and because the chemical is one of the most potent ozone-destroying chemicals affecting the planet's upper-atmosphere ozone shield. Under the Montreal Protocol, methyl bromide use was to have been phased out by 2005, but the U.S. continues to permit its use on certain crops, like conventional strawberry, under so-called "critical use" exemptions.

**Question 19-2:** Research the concept of "critical use" as applied to methyl bromide. Where is its use still permitted in the U.S.? Why?

California's Pajaro River (see Figure 19-2) is listed by the group American Rivers[5] as one of the ten most endangered rivers in America. In the Pajaro Watershed strawberries are grown on fewer than 23,000 acres (9,000 ha) of the region's rich alluvial soil. Yet,

**FIGURE 19-2** Sod farms in the floodplain of the Pajaro River, California, one of the nation's most important agricultural regions. (Courtesy of Jitze Couperus.)

---

[5] www.AmericanRivers.org.

conventional strawberry growers use up to 300 pounds of active pesticide ingredients each year on each acre. (Active ingredients make up only a few percent by weight of pesticides. So the total pesticide amount used on crops is much higher.)

When the Pajaro floods, a toxic brew of pesticides, fungicides, silt, and fertilizer is washed downriver, sometimes into residents' homes but eventually into the Monterey Bay Marine Sanctuary. You can find info on the Pajaro at www.pajarowatershed.com.

**Question 19-3:** The Sierra Club and other environmental organizations calculate that organic, pesticide-free strawberries can be grown for 25 to 30 cents a pound more than conventional strawberries. Would you pay 25 to 30 cents more for a pound of organic strawberries? Why or why not?

The federal government spends up to $30 billion a year on agricultural subsidies, most of it going to mass-producers of "commodity crops" like soybeans, corn, cotton, and wheat. As a result of such subsidies, these crops are usually produced in quantities far exceeding the demand for them.

**Question 19-4:** Do you think subsidizing organic food production rather than commodity crops to reduce pollution, protect soils, and protect public health would constitute a good use of public money? Why or why not?

## AGRICULTURAL PRACTICES THAT REDUCE SOIL LOSS

The loss of topsoil represents one of the greatest threats to agricultural production. (Recall Figure 19-1). About two-thirds of topsoil loss is caused by rain erosion, with another one-third by wind erosion. As we mentioned above, in some regions topsoil is currently being lost up to 300 times faster than it can be replaced. But farmland can be degraded in other ways besides erosion. Mechanical tilling can lead to soil compaction, reducing infiltration capacity, and crusting. Repeated cropping without sufficient fallow (rest) periods or replacement of nutrients with cover crops, manure, or fertilizer can deplete soil nutrients. The loss of topsoil directly affects agricultural productivity (Figure 19-3).

**Question 19-5:** Study Figure 19-3. How much more grain would be produced on 10-inch topsoil compared to 4-inch topsoil?

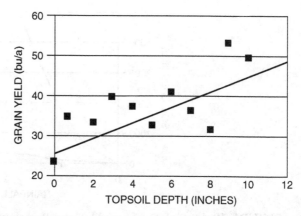

**FIGURE 19-3** Impact of topsoil loss on grain yield. (Havlin, J. L., H. Kok, & W. Wehmueller. 1992. Soil erosion—Productivity relationships for dryland winter wheat, pp. 60–65 in J. L. Havlin (ed.), *Proc. Great Plains Soil Fertility Conf.*, Denver, CO, Mar. 3–4, 1992. Courtesy of J. L. Havlin, H. Kok, and W. Wehmueller/Agricultural Research Service.)

**Question 19-6:** Based on a mid-2012 wheat price of $6.25 per bushel, how much added value per acre does this represent?

Oregon State University's (OSU) Extension Service reports that no till agriculture coupled with abandoning the traditional summer fallow system can substantially limit topsoil erosion and increase soil $CO_2$, whereas conventional agriculture leads to topsoil erosion and reduced soil $CO_2$.

**Question 19-7:** What is the major contrast in $CO_2$ retention between conventional and no-till agriculture? Explain why this might be an important issue.

OSU's Extension Service reports that

> For nearly 100 years, most growers used a summer fallow system. That is, land planted with wheat one year was plowed but left unplanted the following year, to give the soil time to build up enough moisture to produce another crop. But this practice results in massive erosion, such that organic matter in the soil is now one-half its 19th century level.[6]

**Question 19-8:** Study Figure 19-4, which shows how tillage practices influence topsoil loss. How much topsoil loss is avoided during a 1-inch rainstorm by using no-till agriculture compared to the plow-up/down method?

---

[6] http://extension.oregonstate.edu. See for example http://extension.oregonstate.edu/catalog/html/sr/sr1083-e/ sr1083_08.pdf.

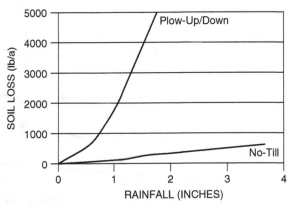

**FIGURE 19-4** Soil loss reduced by no-till agriculture, Columbia basin, Oregon. (Dickey, E., P. Harlan, & D. Vokal. 1981. Crop residue management for water erosion control. NebGuide G-81-554. University of Nebraska-Lincoln, NE. Courtesy of University of Nebraska-Lincoln.)

In the early 1990s, the Columbia Basin had 625,000 acres in summer fallow annually. Annual cropping was virtually nil. By 2000, 16 percent of the traditional summer fallow acreage—more than 100,000 acres—was being planted every year, and this number was increasing. Planting a crop each year reduces soil erosion. On high-yielding wheat fields, it cuts soil erosion from 12 tons per acre under the fallow system to 6 tons.

Growing crops every year could offer growers a product other than wheat to sell in more favorable domestic and niche export markets. This could help to reduce or eliminate subsidy payments.

Even more important, however, is the future of farming in the Columbia Basin. Annual cropping provides an alternative to a wasteful farming practice that will, if continued, eventually put farmers out of business in the Columbia Basin.

**Question 19-9:** Should the government require farmers like those in the Columbia Basin to use no-till methods or lose government crop subsidies? Why or why not?

# CAN AID STOP SOIL DEGRADATION?

According to scientists at the International Food Policy Research Institute (IFPRI),[7] soil degradation has already had significant impacts on the productivity of about 16 percent of the globe's agricultural land. Their study concludes that almost 75 percent of crop land in Central America is seriously degraded, 20 percent in Africa (mostly pasture), and 11 percent in Asia.

**Question 19-10:** Desertification creates millions of environmental refugees. How could such land degradation affect U.S. population growth?

---

[7] www.ifpri.org.

Threats to the world's food production capacity are compounded by three disturbing trends, according to experts at the IFPRI:[8]

1. One billion more people could be on the planet by 2030, almost all in poorer developing countries.
2. The natural fertility of agricultural soils is generally declining.
3. It is increasingly difficult to find productive new land to expand the agricultural base.

One of the most common management techniques used to maintain the productivity of farms is the use of inorganic fertilizers (nitrogen, phosphorus, and potassium) or manure. Too little can lead to soil "nutrient mining" (amount of nutrients extracted by harvested crops is greater than the amount of nutrients applied), and too much can lead to nutrient leaching (washing away of excess nutrients contaminating groundwater and surface water). Nutrient depletion can result in severe water pollution.

IFPRI experts have found that African farmers apply only one-fifth the necessary fertilizers to replace nutrients lost each year. But simply providing fertilizer aid to farmers is only part of the solution and is more complex than it sounds. For example, *Scientific American* reports, "it costs more to transport fertilizers 100 kilometers inland from a port anywhere in sub-Saharan Africa than it does to ship them to that same port from North America."[9] This is partly due to poor roads and other infrastructure in Sub-Saharan Africa.

**Question 19-11:** Road construction paid for by international agencies like the World Bank has been criticized by environmentalists, who cite its potential for ecosystem destruction. Do you support using U.S. aid for road building? If not, can you suggest alternatives to road construction to alleviate the fertilizer shortage in Sub-Saharan Africa?

# SOILS AND CO$_2$

Soils store $CO_2$. While the burning of fossil fuels is the major source of anthropogenic atmospheric $CO_2$, soils may be a significant source as well. Soil microbes and other organisms break down the humus—organic matter—from leaves and other sources and release $CO_2$. Scientists know that, without human interference, soils would add an amount of $CO_2$ to the air equal to that amount added to the soil from plants. But that may no longer be the case.

According to the American Geological Institute, world soils contain about 1,100 to 1,600 billion metric tons (bmt) carbon, more than twice the carbon in living vegetation (560 bmt) or in the atmosphere (750 bmt).[10] Therefore, even small changes in soil carbon sequestration per unit volume could have a major impact on the global carbon balance.

Scientists now believe that atmospheric $CO_2$ started to increase during the nineteenth century, before the intensive use of fossil fuels. The nineteenth century increase in atmospheric $CO_2$ was likely derived from continent-wide conversions of woodland and prairie

---

[8] Ibid.

[9] www.sciam.com/article.cfm?articleID=00084D86-A5FD-1488-A5FD83414B7F0000.

[10] www.agiweb.org, and see http://www.geotimes.org/jan02/feature_carbon.html.

to agriculture in Europe and North America. The stripping of the native ground cover caused the rate of $CO_2$ emission from soils to increase rapidly, resulting from a higher rate of humus decomposition. This increase in $CO_2$, in turn, contributed to an 0.8°F increase in temperature over the past hundred years. And this growing global temperature accelerated the decomposition of humus, a classic example of "positive feedback."

Moreover, modern industrial agriculture, through poor farming practices leading to soil erosion, removes more organic matter from the soil than it replaces, resulting in a growing imbalance between organic matter added to soils and $CO_2$ emitted to the atmosphere from decomposition.

As the Earth's temperature rises, each 0.9°F increase in global temperature will add an amount of $CO_2$ from the soil equal to a year's burning of fossil fuels, according to researchers at the University of California-Irvine.

## SUMMARY: CAUSES FOR CONCERN

Agriculture must provide an adequate quantity of decent food at affordable prices, a challenge that farmers and scientists in developed countries have met with considerable success. Global food prices dropped by nearly 40 percent between 1970 and 2000 because of crop yield increases, investments in agricultural research, and *public and environmental subsidies*. However, since 2000, to take one example, increasing use of corn (maize) in ethanol production has raised the price.

Over the past three decades, the per capita increase in production of the world's three major cereal grains has been 37 percent for maize, 20 percent for rice, and 15 percent for wheat. Lowering the prices of major staples can directly benefit the poor, who spend a large part of their income on purchasing food, but falling farm prices also put farmers out of business and lead to high levels of waste-producing and distorting government subsidies.

These upward trends, as we have seen, come with an environmental cost—soil degradation—that has not been reflected in the price of the food. Remember that commodity prices do not include the cost of degraded land or ecosystems. That cost has been borne by the environment.

The unprecedented scale of agricultural expansion and intensification during the last half of the twentieth century raises two concerns, as expressed by the World Resources Institute:[11]

> *First,* there is growing concern over the vulnerability of agricultural lands to the stresses imposed on them by the intensification of agriculture. Can technology and more fertilizer continue to offset the depletion of soil fertility and freshwater resources (see Issue 11)? As soil fertility declines and water becomes scarcer, what will be the impact on food prices?
>
> *Second* are concerns about the *externalities* of agricultural production that are often made worse by intensification. These negative impacts include additional stresses that agricultural land can cause beyond their own boundaries (air and water pollution, dust, spread of genetically modified crops, etc.) but that are not properly reflected in production costs nor in the prices consumers pay for agricultural products.

At a watershed level, these impacts include the following:

- Reduced river flows and groundwater levels
- Increased soil erosion, affecting fisheries and riverine infrastructure (canals, dams, etc.)
- Damage to both aquatic ecosystems and human health arising from fertilizer and pesticide contamination in water or on crops

---

[11] www.wri.org, and see http://insights.wri.org/news/2011/09/new-approach-feeding-world.

■ Loss of habitat and biodiversity from converting land to agriculture, as well as the declining genetic diversity of domesticated plant and animal species currently in use.

Among issues raised by the World Resources Institute are the following:

■ While U.S. restaurants throw away vast quantities of food (35 million tons of food scraps in 2010) one billion people in the world's poorest countries cannot afford to purchase sufficient food.
■ Most global agricultural production, with the exception of dairy and perishable vegetable production, still comes from intensively managed industrial sites located far from population centers. This makes distribution of agricultural products dependent on fuel prices and roads.
■ While the global expansion of agricultural land has been modest in recent decades, intensification has been swift. Irrigated area grew by more than 70 percent over the past thirty years, stressing many critical groundwater resources and resulting in the disappearance of many water bodies, like the Aral Sea. Even though irrigated lands account for only 5.4 percent of agricultural area globally, they reach 35 percent in South Asia, 15 percent in East Asia, and 7 percent in South East Asia.

**Question 19-12:** Research advantages and disadvantages of the organic and "locavore" food movements. Comment.

Protecting soils and agricultural systems and ensuring adequate water supplies are among the greatest challenges facing humanity in the twenty-first century.

**Question 19-13:** Discuss whether the current level of soil degradation and erosion is consistent with sustainability and sustainable societies.

**Question 19-14:** Summarize the main points of this Issue.

## FOR FURTHER THOUGHT

**Question 19-15:** A significant fraction of atmospheric carbon dioxide, the most common greenhouse gas, comes from burning tropical forests, much of which were cleared to raise cattle. These cattle, in turn, release methane, a more potent greenhouse gas. Research how raising cattle contributes to global warming. Summarize your findings in a paragraph or two.

**Question 19-16:** Go to the websites www.chesbay.org and www.cbf.org, and research the contamination of Chesapeake Bay by fertilizer, pesticide, silt, and manure from agricultural runoff. Some researchers call the Bay an ecological disaster. Why? Identify solutions proposed by the organization. Who will pay for them?

**Question 19-17:** The Aral Sea was essentially destroyed in the 1950s by the former Soviet Union to provide irrigation water for cotton. Research the impacts of the diversion of the rivers that formerly nourished the Aral Sea. How are these effects being addressed?

**Question 19-18:** Is it sustainable to encourage or subsidize agricultural practices which result in artificially cheap food, made artificially cheap because much of the real cost of production is borne by the environment? Do you think these practices can continue indefinitely? Explain your answers.

**Question 19-19:** Research the issue of carbon sequestration in soils. What methods have been proposed to retain carbon in soils?

**Question 19-20:** Could conversion to organic agricultural methods worldwide halt degradation and provide an adequate food supply for humanity?

# Issue 20

## THE STATE OF GLOBAL FORESTS

## KEY QUESTIONS

- Why are forests important?
- Where do we need forested areas to protect biodiversity?
- How do present forests compare to prehistoric forests?
- What are trends in changes in forest cover?
- How do tropical forests, temperate forests, and boreal forests differ?
- Can forests help mitigate climate change?
- How do consumer practices threaten global forests?
- What practices can help restore and protect forests?

## INTRODUCTION

Most people know that we are rapidly losing tropical forests. A study by the World Resources Institute[1] (WRI) concluded that over half of the original planetary forest cover, or about 3 billion hectares, has been lost due to human activity, and we continue to lose forest at a rate of about 16 million hectares a year. For comparison, the total forest area of Canada is about 418 million hectares.

Forty percent of the planet's remaining forests are frontier forests. Frontier forests, as defined by WRI, have large contiguous areas with limited human influence and can sustain biodiversity levels without human interference. Other kinds of modern forests, such as pine monocultures, have undergone so much human alteration that they have diversity levels much reduced from prehistoric times. Large predators, for example, require extensive tracts of contiguous habitat and can't survive in fragmented forest.

Almost half of the world's forest has been replaced by agriculture, pasture, or settlement over the past 8,000 years. From 1850 to 1990 alone, deforestation released over 120 billion metric tons of carbon into the atmosphere, initiating the present episode of accelerated global climate change.

**Question 20-1:** What was the historic rate of forest loss over this 8,000-year interval, in percent per year?

---

[1] www.wri.org.

According to the Nature Conservancy, "[As of 2005] every second, a slice of rainforest the size of a football field is mowed down. That's more than 56,000 square miles of natural forest lost each year."[2]

**Question 20-2:** Based on an original global forest area of 6 billion hectares, what is the rate of forest loss per year, as of 2005?

In Brazil alone between 2000 and 2005, 3.1 million hectares of forest were cut down per year. However, due partly to government regulations and partly to a drop in worldwide soybean demand, the deforestation rate in Brazil fell by 75% between 2004 and 2012.[3]

**Question 20-3:** Compare the total forest loss in Brazil during 2000-2005 to Canada's total forest.

For Indonesia the deforestation rate was 1.85 million hectares per year.

**Question 20-4:** The U.S. had about 750 million acres of forest in 2000. Compare the total forest loss in Indonesia between 2000–2005 to the total U.S. forest cover.

# WHY PROTECT FORESTS?

Why is it important that we restore and preserve intact forest ecosystems?

■ Even fragmented forests help maintain biodiversity.
■ Forests store immense amounts of carbon and thus help moderate the greenhouse effect and global climate change.
■ Forests provide billions of dollars worth of economic services, such as wood products, herbs, medicines, and other raw materials.
■ Forests can protect water supplies, clean the air of pollutants, moderate severe weather like hurricanes, and provide recreational facilities and solitude.
■ Because forests can generate precipitation and can even create their own climate, as in the rainforests of the tropics.
■ Because at least 60 million people live in intact forests and depend on them for a livelihood.
■ Because it is the right thing to do, in terms of our responsibilities to future generations.

[2] The Nature Conservancy, http://www.nature.org/ourinitiatives/urgentissues/rainforests/rainforests-facts.xml.
[3] *Science*, http://news.sciencemag.org/scienceinsider/2010/09/deforestation-rate-continues-to.html.

**FIGURE 20-1**  Recently developed strip mall along Highway 17 bypass in Pawleys Island, South Carolina. The site had been a small pine forest prior to construction of the store "Primarily Pine" and others. (D. Abel)

**Question 20-5:**  Figure 20-1 shows a strip shopping center in Pawleys Island, SC. The site (including the land occupied by the store "Primarily Pine") was a small pine ecosystem before the land was completely cleared. In similar cases, developers have responded to criticism about their land-clearing practices by arguing that the forest was in fact fragmented and already disturbed (i.e., not pristine), that we must balance the economy and the environment, and/or the land was private property and the developer had the right to develop it. Evaluate the developers' responses. Discuss whether you agree or disagree.

# THREE CLASSES OF GLOBAL FORESTS

There are three types of global forest communities, based roughly on latitude, but defined by seasonality:

- *Boreal* forests, at polar latitudes and high elevations anywhere
- *Temperate* forests, at mid-latitudes
- *Tropical* forests, at frost-free equatorial latitudes

## Tropical Forests

Figure 20-2 shows the global extent of tropical rainforest. Daylight is roughly constant year round, there is no frost, and precipitation is heavy. Soils are generally poor, easily depleted of nutrients, and tend to be acidic, making large-scale western-style agriculture difficult. Precipitation often exceeds 200 centimeters per year. Flora is very diverse. One hectare may contain dozens of plant species.

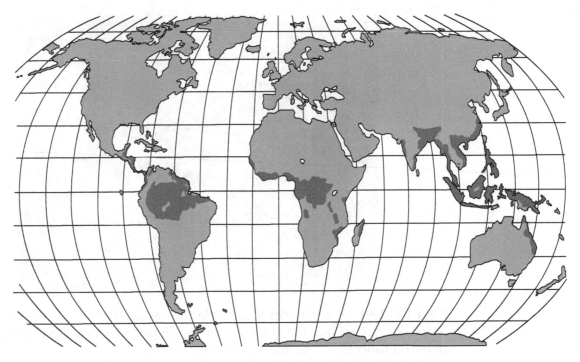

**FIGURE 20-2** The extent of global tropical rainforest (darkened areas).

Alarmist rhetoric can perhaps be anticipated from some environmental activists. However, here is the U.S. National Aeronautics and Space Administration (NASA):[4]

> The loss of tropical forest is more profound than mere destruction of beautiful areas. If the current rate of deforestation continues, the world's rainforests will vanish within 100 years, causing unknown effects on global climate and eliminating the majority of plant and animal species on the planet.

**Question 20-6:** Identify the effects of population growth on tropical deforestation, both in the home country and abroad.

Deforestation is occurring for a number of reasons:

■ Commercial logging
■ Livestock raising, mainly cattle
■ Food and fuel crops, both subsistence and commercial

Some deforestation is driven by debt. Some countries owe billions of dollars to foreign lenders and must either pay interest on that debt or default. Commercial logging provides revenue to pay foreign lenders. Deforestation may also be driven by poverty and ignorance. In Bolivia, which controls part of the Amazon rainforest, annual GDP is about $4800 per capita (it was $48,000 for the U.S. in 2011) and 51% of the populace lives on less than the

---

[4] NASA: www.earthobservatory.nasa.gov/Library/Deforestation.

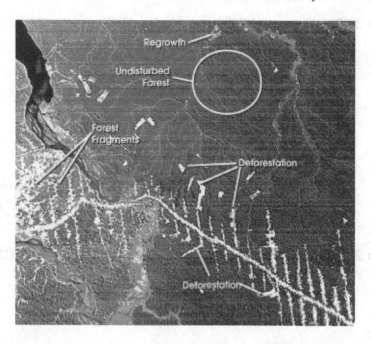

**FIGURE 20-3** Deforestation in the Brazilian rainforest. (NASA)

equivalent of $2 (U.S.) per day.[5] Mines and hydroelectric dams, often built using foreign loans, may destroy, contaminate, or flood thousands of hectares of rainforest. Roads and settlements also contribute to deforestation, roads perhaps most importantly, since roads provide access to miners, loggers, and desperate farmers. Figure 20-3, a part of the Brazilian rainforest, shows how deforestation commonly occurs. This pattern generates isolated areas called "forest fragments," in which plant and animal species are disturbed. Alien species are commonly introduced along roads by vehicles. Humans using roads may then harvest animals in the forest for "bush" meat, one of the fastest growing sources of protein in equatorial Africa.

***Bush meat, logging, and deforestation in African rainforests.*** Killing of forest animals including endangered primates for food has become epidemic in African rainforests. There are four key concepts involved in understanding and addressing this issue.[6]

1. The bush meat "crisis" is the most immediate threat to wildlife populations in Africa.
2. Human population growth, illegal hunting methods, logging, and road building are essential components of the crisis.
3. Forests over much of west and central Africa are literally being emptied of their wildlife by illegal hunting.
4. Solutions to this problem require international action.

The economics of the African bush meat issue are sobering. More than 30 million people live in central African forests, and they eat approximately 30 kilograms of bush meat, mainly large mammals, per capita annually as of 2009.

---

[5] CIA World Factbook, www.cia.gov.
[6] www.bushmeat.org.

**Question 20-7:** How much bush meat in total is "harvested" from central African rainforests annually, in metric tonnes?

Consider the Democratic Republic of the Congo, where the 2011 per capita income was around $300, ranking 226th (last) on the U.S. CIA's global list.[7] A bush meat hunter can make more than $300 annually hunting and selling bush meat. The population of 73.6 million people (2012) is growing at 3 percent per year.

**Question 20-8:** What is the doubling time for this population? (Use t = 70/r; see "Using Math in Environmental Issues," pages 6–8.)

**Question 20-9:** What is your long-term forecast for the survival of the Congolese rainforest and its megafauna should these population projections prove accurate?

## Temperate Forests

Temperate forests, at mid-latitudes of both the northern and southern hemispheres, have well-defined seasons with several frost-free months. Flora is less diverse than in tropical forests, with fewer tree species per hectare.

However, in the United States, due to fire suppression techniques applied beginning in the 19th century, many temperate forests have "too many trees." For example, research suggests that pre-colonial New Mexico temperate forests had perhaps a dozen trees per acre. Now, there may be ten times as many, leading to disastrous fires. This overabundance of trees has another unintended consequence: much of the precipitation is returned to the atmosphere by the thick tree canopy, depriving the soil of much-needed moisture.[8]

Varieties of temperate forests include southeastern (U.S.) pine forests, Pacific Northwest conifer-dominated rainforests, and Mediterranean forests with rainy winters and very dry summers. Little original temperate forest remains on Earth. For example, less than 1 percent of Europe's original temperate forest remains.

---

[7] CIA op cit.

[8] *The Economist.* http://www.economist.com/blogs/babbage/2012/06/forest-management.

One distinctive characteristic of temperate forests is the size of its largest trees. A Douglas fir in Washington's Olympic Peninsula was 420 feet tall when chopped down in 1895. Some redwoods in California forests stand over 300 feet tall.

One-quarter of the Earth's remaining temperate rainforest is in British Columbia, Canada, and adjacent southeastern Alaska. Over half of these forests have already disappeared. It is home to at least 3,000 distinct wild Pacific salmon runs. Its integrity, if not its very survival, is threatened by clear-cut industrial logging.

Canada's 30,000-member David Suzuki Foundation has undertaken a scientific analysis of the needs of this ecosystem.[9] It has concluded that logging but not industrial clear-cut logging can be compatible with ecosystem integrity. It has proposed a set of principles to guide logging activity in this irreplaceable resource. These principles could well be applied to almost any type of natural resource extraction activity. They are as follows:

1. The fundamental goal of management should be ecosystem health, not economic gain.
2. All local activities should be consistent with protecting ecosystem integrity.
3. The cut rate must be low enough to sustain the integrity of the entire ecosystem.
4. Indigenous people and other local stakeholders' views must be central to planning and decision making.
5. *All* native plant and animal species must be preserved.
6. Waterbodies—the adjacent marine environment and all rivers, streams, wetlands, and lakes—are vital for the health of the ecosystem and must be protected without degradation.
7. Determining what to retain is more important than what to remove. The view of some loggers that much vegetation in forests is "trash"—"trash trees," for example—is a view borne out of ignorance and has no place in ecosystem management.
8. Ecological restoration must be incorporated into decision making at all levels.
9. Apply the precautionary principle (see page 14) in all decisions affecting logging activity.

**Question 20-10:** Select three of the nine principles and critique them using a critical thinking approach (pages 8–12). Are there any of the entire list that you would delete? Modify? Are there any that you would add? Explain.

## Southern Forests

In the early 1600s, as much as 350 million acres of undisturbed forests, an area nearly the size of Texas, California, and Montana combined, blanketed the southern U.S. Virtually none of this remains pristine: 99% of southern forests have been cut in the last 400 years. The majority of southern forests are *commercial* forests, typically meaning rows of loblolly pines or other fast-growing trees scattered among recently harvested clear cuts. Known as *tree plantations,* these areas may superficially resemble forests, but they are *monocultures* (composed of a single kind of tree), their biodiversity is not as high as that of undisturbed,

---

[9] David Suzuki Foundation. www.davidsuzuki.org.

natural forests, and they do not provide the same levels of ecosystem services (e.g., erosion protection, storage of carbon dioxide) that natural forests provide and on which humans depend.

Natural forests in the southern United States contain some of the most biologically rich ecosystems in North America. Many of the region's plant and aquatic species can be found nowhere else in the world. According to the Dogwood Alliance's Scot Quaranda, Southern forests contain[10]

■ The highest concentration of tree species diversity in North America;
■ The highest concentration of aquatic diversity in the continental United States, including the richest temperate freshwater ecosystem in the world; and
■ The highest concentration of wetlands in the United States, 75 percent of which are forested.

"Nowhere in America is there a greater variety of native plant communities, native plant species, or rare and endemic plants."[11]

Here are more facts about southern forests:

■ Approximately 6 million acres of the South's forests are logged every year, largely to make paper.
■ Removals of softwoods (pines) exceed growth throughout the region.
■ Removals of all species exceed growth on industry land in the region.
■ Logging is expected to increase 50 percent by the year 2040.
■ Removals of hardwoods are projected to exceed growth by 2025.

A 2010 study, *Quantification of Global Gross Forest Cover Loss*, in the journal *Proceedings of the National Academy of Sciences*, reports that forest loss in the southeast U.S. is among the highest globally.[12] According to the U.S. Forest Service, the South produces more timber than any other single *country* in the world. It is projected to be the dominant producing region for many decades to come. The South accounts for 15% of the world's paper supply. This industry is driving the destruction of our forests through industrial-scale clear-cutting, and the conversion of forests and wetlands to intensively managed tree plantation monocultures. As of 2009, 43 percent of southern forests were "sterile pine plantations" requiring millions of pounds of pesticide applications each year. [13]

An emerging threat to southern forests is logging for biomass as an energy source. *Biomass* is a catchall term for a variety of energy sources derived from organic matter, including high-yield grasses (e.g., switchgrass), food crops (corn, sugarcane), landfill methane, food waste, and wood (also known as *forest* biomass). Biomass is generally considered an attractive alternative to fossil fuels because it promotes energy security, decreases foreign oil dependence, contributes to a green economy, and is thought to contribute less to climate change. Increasing demand for forest biomass for domestic use and export to Europe is changing the nature of forest biomass use in the southeast from low-impact, small-scale, Mom-and-Pop operations that rely on waste from sawmills and other logging residues, to an industrial scale that demands harvest of large stands of whole, standing trees.

This shift in turn has led to increased scientific scrutiny of the environmental impact, particularly the key assertion that the use of forest biomass is *climate neutral*. Climate neutrality means that combusting wood releases carbon dioxide that will once again be stored in the forest when it regrows, and that this cycle will result in no net addition of carbon dioxide to the atmosphere over long periods.

[10] Dogwood Alliance, www.dogwoodalliance.org and personal communication.
[11] U.S. Forest Service (USFS).
[12] Available at http://www.pnas.org/content/107/19/8650.full.
[13] USFS, ibid.

A recent report commissioned by the National Wildlife Federation (NWF) and Southern Environmental Law Center (SELC), *Biomass Supply and Carbon Accounting for Southeastern Forests,*[14] rigorously analyzed the climate change assumption. The study's conclusion: Yes, burning forest biomass recycles carbon dioxide over a period from 35–50 years, after which there is a net benefit, but in the short term there is a net increase in carbon dioxide released into the atmosphere so that any long-term benefit will be too late to avoid the worst impacts of climate change. Other scientific studies have calculated that carbon dioxide balance may take a century or more. Moreover, mature, natural forests are better storehouses of carbon dioxide than tree farms.

**Question 20-11:** Propose ways to reduce the destruction of southern forest ecosystems.

## Boreal Forests

Boreal forests, also called "taiga," cover over 1 billion acres of Siberia, Scandinavia, and northern Canada.[15] Found roughly at 50 to 60 degrees north latitude, boreal forests have nutrient-poor, acidic soils and are characterized by short growing seasons and severe winter conditions. *Permafrost* characterizes much of the boreal realm. This permanently frozen soil may store huge quantities of methane. Boreal forests are dominated by hardy coniferous varieties, like spruce and fir.

Although most popular focus is on threats to tropical forests, intensive logging in boreal forests may result in their disappearance in one or two generations without careful management. The boreal region may contain up to one-third of terrestrial carbon, according to the Polaris Project.[16] To what extent intact boreal forests survive will depend on actions taken by two countries: Russia and Canada.

*Russian boreal forests.* Russia has the largest mass of boreal forest on Earth but little effective protection of its remaining boreal forests, outside of a few small nature areas and national parks. Low prices for lumber in Russia compared to adjacent countries like Finland contribute to indiscriminate forest loss. Deforestation in Russia may reach 20,000 km$^2$ (1 km$^2$ = 100 ha) annually, similar to the annual deforestation rate in the Amazon Basin of Brazil. Here is one summary of the importance of Russian forests from the Woods Hole Research Center.[17]

Russian forests contain approximately 56.3 billion tonnes of carbon (bmtC) in vegetation, and about 135.7 bmtC in soil organic matter. Earth's circumpolar boreal forests and associated bogs (peatlands) contain at least five times the carbon of the world's temperate forests, and twice the carbon in tropical forests. Approximately 60% of the C is presently in permafrost. In short, Russia holds nearly half of the Northern hemisphere's terrestrial carbon. Thus, "Russia's natural forest resources play an integral role in global carbon cycling and climate change."

---

[14] Available at: http://www.southernenvironment.org/uploads/publications/biomass-carbon-study-FINAL.pdf.

[15] World resources Institute. www.wri.org.

[16] Polaris Project, www.polarisproject.org.

[17] www.whrc.org.

**Question 20-12:** Would you be in favor of a global fee or surcharge on paper or wood products to help protect Russian boreal forests? Why or why not?

**Question 20-13:** Should protecting global carbon stores be entirely a national (in this case, Russia's) responsibility, given the integration of global economies? Comment, and suggest alternatives.

*Canadian boreal forests* The Canadian boreal forest is the single largest ecosystem on the North American continent. It makes up more than a third of Canada, yet less than 3 percent is protected from industrial logging, mining (including oil and gas development), or hydro development, the three greatest threats to the boreal forest. As of 2006, Canada was one of the world's largest exporters of forest products, exporting $39 billion ($C) and the United States was by far its biggest customer. However, China may overtake the United States in the global marketplace, and Russia could be that nation's supplier of choice.

Oil and gas exploration is a little-appreciated threat to the integrity of Canada's forest, through the laying of many thousands of kilometers of seismic lines, which require cutting a swath through the forest.

As to the future, here are the findings of a report prepared by the Boreal Forest Network:[18]

- Canada's forests are managed mainly for timber, but the public values the forests mainly for nontimber use.
- Nearly a third of the forest is within one kilometer of a road.
- Half of Canada's boreal forest is owned by a few large timber companies.
- At least 300 hydroelectric dams and sixty large, active mines operate within the boreal zone.
- About 1% of Canadian boreal forest is logged each year, and the preferred method is clearcutting.
- 90% of logging in Canada occurs within primary and old-growth forests—forests of high biodiversity and wilderness value.

The Boreal Forest Network concludes,

(S)olutions to protect . . . the boreal forest of North America must first include a coordinated campaign in both the U.S., as the primary consumer nation, and Canada. This campaign would be designed . . . to influence both Canadian and U.S. Consumers . . . about the underlying causes of deforestation and the social and economic impacts it has had on communities living in the boreal eco-zone, as well as the global impacts associated with current large-scale resource extraction practices.[19]

---

[18] www.borealnet.org.
[19] Ibid.

**Question 20-14:** What responsibilities, if any, do American consumers have to protect Canadian boreal forests? What form could such activity take?

## GLOBAL FORESTS AND JUNK MAIL

Junk mail is the popular term for what the U.S. Postal Service calls "standard" mail. It consists mainly of unsolicited advertising circulars and mailings. While Americans receive about 5 million tons of junk mail a year, the Direct Marketing Association reports that nearly half of junk mail is discarded unread. Moreover, it costs localities $1 billion each year to dispose of junk mail. Seventeen trees are needed to make a ton of paper. Six and a half million tons of paper are used each year to produce junk mail.[20]

**Question 20-15:** There are about 310 million Americans. How many pounds of junk mail on average does each American receive?

**Question 20-16:** Since at least 44% of junk mail is discarded unread, how many trees are thrown away each year in unread unsolicited junk mail?

## GLOBAL FORESTS AND DISPOSABLE PAPER CUPS

U.S. consumers used 16 billion paper cups in 2006.

> If only 50 customers a day in every store were to use reusable mugs, Starbucks would save 150,000 disposable paper cups daily. This equals 1.7 million pounds of paper, 3.7 million pounds of solid waste, and 150,000 trees a year.[21]

Each Starbucks paper cup weighs on average almost 0.7 ounce. That's about 24 paper cups to the pound.

Dan Welch, owner of Portland Oregon's World Cup Coffee, said that the average 16-ounce paper cup and cardboard sleeve, including artwork, plastic lid, and stir stick, costs more than 22 cents to produce.[22] This cost, of course, does not include disposal. Starbucks

---

[20] www.donotmail.org.
[21] Starbucks Inc, www.starbucks.com.
[22] www.portlandmercury.com.

is, as one observer said, merely "the tip of the iceberg." Fast-food giants like McDonald's, Burger King, Wendy's, 7-11, and Jack-in-the-Box are responsible for far more disposable cup waste than the coffee shop chain. For example, McDonald's[23] used more than 451,000 tons of packaging in 2003, a 12 percent increase over 2002. About 41 percent was paperboard, the raw material for paper cups.

**Question 20-16:**  How much paperboard did McDonald's use in 2003?

**Question 20-17:**  Using the value 24 cups per pound and 16 billion cups used in 2006, how many pounds of paperboard did this represent?

**Question 20-18:**  Given these glimpses into the amount of wood products used in paper cups and junk mail, assess the importance of education and small lifestyle changes in protecting global forests.

**Question 20-19:**  Is your current use of forest products consistent with the goals of sustainability? In what ways might you alter your use of forest products to slow the destruction of forests globally?

**Question 20-20:**  Summarize the major points of this Issue.

---

[23] McDonald's Corp., http://www.aboutmcdonalds.com/mcd/sustainability/library/policies_programs/sustainable_supply_chain/Environmental_Scorecard.html

# FOR FURTHER THOUGHT

**Question 20-21:** Go to www.donotmail.org for ways to stop junk mail. Do you think advice on receiving less junk mail belongs in an issue on the state of global forests? Why or why not?

**Question 20-22:** Some irate victims of junk mail return the business reply envelope (BRE) or card enclosed with the junk mail to the sender. This has two advantages: it costs the sending company money and requires them, or their agent, to deal with the junk mail. It also provides revenue to the Postal Service. It will not, however, end solicitations because companies usually contract out opening and processing BREs to third parties, who only get paid by returning bona fide purchase requests to the company. However, if enough people returned BREs empty or with the original advertising material returned, the resultant cost might make them think twice about using junk mail.

Is returning BREs filled with junk mail a suitable way to inform advertisers of your irritation? Explain.

**Question 20-23:** The environmental group ForestEthics seeks to save boreal forests through a campaign that targets corporations that use products made from boreal forest trees. Go to www.forestethics.org and evaluate the list of corporations and what they are doing to reduce environmental impacts of their products.

**Question 20-24:** In 1993, in response to high rates of global deforestation, the Forest Stewardship Council (FSC) was formed to promote responsible, sustainable management of forests. For example, FSC certification bans toxic chemicals and does not allow forest clearcuts. In addition, for wood to achieve FSC certification, it must avoid these wood origins:

1. Illegally harvested wood

2. Wood harvested in violation of traditional and civil rights

3. Wood harvested in forests in which High Conservation Values (areas particularly worth of protection) are threatened through management activities

4. Wood harvested from conversion of natural forests

5. Wood harvested from areas where genetically modified trees are planted

In 1994, the forest industry launched it's own certification, the Sustainable Forestry Initiative (SFI). According to its website, "SFI forest certification promotes responsible forestry practices. . . . It is based on principles and measures that promote sustainable forest management and consider all forest values. Critics argue that SFI does not promote sustainable forestry in that it allows: large clearcuts, logging in sensitive areas, toxic chemicals, genetically-modified trees, and conversion of old-growth forests to tree plantations.

FSC certification is considered the gold standard for responsible, sustainable forestry, and is widely supported by the environmental community. Why then do you think the forestry industry devised a new standard? Research the issue. Which certification maximizes forest sustainability?

# RESTORING ESTUARIES:
## *CHESAPEAKE BAY*

## KEY QUESTIONS

- What are the sources and the impacts of pollution on Chesapeake Bay and its watershed?
- What is the status of remediation efforts?
- What actions are being taken to reduce the environmental impact of population growth there?
- What is the long-term prognosis for Chesapeake Bay?

## BACKGROUND

The Chesapeake Bay formed about 8,000 years ago as rising sea level drowned the mouth of the Susquehanna River. It is the nation's largest and most productive estuary.[1] There are three interconnected parts of Chesapeake Bay: the Bay proper, the Bay's watershed[2] (Figure 21-1), and the Bay's airshed[3] (Figure 21-2).

The watershed's 64,000 mi[2] area includes parts of six states, the District of Columbia, and 1653 local governments. The watershed had a population around 17 million at the end of 2012, growing at around 1.2 percent per year.[4] Surveys have shown that most residents of the Bay's watershed do not even realize that they live within the watershed. To increase public awareness, in 1997 the states in the watershed put watershed boundary signs along major highways.

The airshed measures 418,000 square miles, or roughly six and a half times the size of the Bay's watershed.

**Question 21-1:** What is the doubling time (see "Using Math in Environmental Issues," pages 6–8) of the watershed's population?

---

[1] An estuary is a coastal embayment where freshwater from rivers and groundwater mixes with salt water.
[2] The watershed is the area drained by streams that feed the Bay.
[3] The Bay's airshed is the geographic area that is the source of any airborne pollutants that can affect the Bay.
[4] www.census.gov.

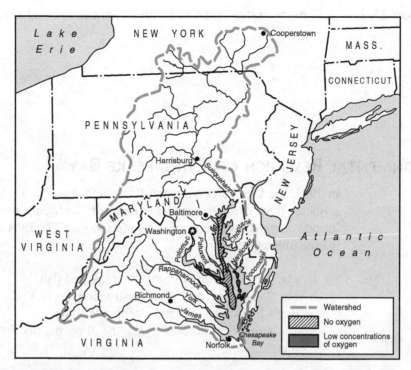

**FIGURE 21-1** Chesapeake Bay and its watershed. (From Blatt, H. 1997. *Our Geologic Environment.* Upper Saddle River, NJ: Prentice Hall. Courtesy of Harvey Blatt.)

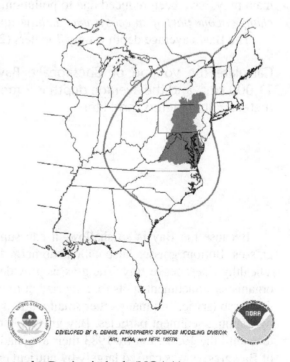

**FIGURE 21-2** Chesapeake Bay's airshed. (NOAA)

**Question 21-2:** How many states and Canadian Provinces are included in the Bay's airshed?

## ENVIRONMENTAL RESEARCH ON CHESAPEAKE BAY

In 1983, the U.S. Environmental Protection Agency (EPA) concluded that the Bay was seriously threatened by *nutrient enrichment*. Nutrients are essential for plant growth and include compounds of nitrogen and phosphorus. The EPA also expressed concern regarding the following:[5]

1. The overharvesting of oysters, crabs, and fish,
2. The 2,000 dams and other obstructions to fish passage that had been built on the Bay's tributaries over three centuries, and
3. The toxic emissions entering the Bay, mainly from industrial and commercial sources.

By 1991, the Bay's once thriving oyster population had collapsed. However, a ban on rockfish (striped bass) harvesting allowed that species to begin recovery. This recovery, however, created another problem: the decline of prey species. Bay anchovies are suffering from overpredation due to the increase in the numbers of striped bass, which are feeding heavily on the bay anchovy. Why? Because populations of Atlantic menhaden, the striper's main prey, have been reduced due to pollution, dams, and overfishing. As you can see, a *change in one part of an ecosystem can have unexpected changes throughout.*

The Bay's average depth is only 7 meters (21').

**Question 21-3:** Calculate the volume of Chesapeake Bay. The Bay's surface area is roughly 11,000 km², and the average depth is 7 meters. Express your answer in cubic kilometers, cubic meters, and liters.

Because the Bay is so shallow, it can support luxuriant bottom vegetation, mainly grasses. Bottom grasses, also called submerged aquatic vegetation (SAV) are one key to a healthy Chesapeake Bay. The grasses provide oxygen for the water, food for some Bay organisms, attachment spots for a myriad of tiny organisms, and hiding places for juvenile fish, crab larvae, and many other small animals. These grasses also promote water clarity by trapping sediment particles. Bay grasses had covered most of the Bay's shallow bottom until the 1950s, but by 1983 their area had been reduced by 90 percent. The decline of Bay grasses was caused largely by nutrient enrichment. Excess nutrients allow floating microscopic algae to flourish, blocking sunlight to the bottom grasses. The surface algae grow rapidly, quickly deplete the nutrients, die, and fall to the bottom. There, their decay

[5] McConnell, R.L. 1995. The human population carrying capacity of the Chesapeake Bay watershed: A preliminary analysis. *Population and Environment, 16* (4): 335–351.

depletes the water of oxygen, resulting in hypoxic (low oxygen) or anoxic (no oxygen) conditions (refer back to Figure 21-1). Since most bottom-dwelling animals cannot live without oxygen, they die or leave, and the bottom, now devoid of plant cover and depleted of oxygen, becomes a veritable desert.

Without Bay grasses to replenish the depleted oxygen, the bottom becomes inhospitable to most animal life. Cyanobacteria and diatoms may replace green algae, for example (for details, see Cooper and Brush[6]).

In 1987, the states of Maryland, Virginia, and Pennsylvania joined with the District of Columbia and a number of federal agencies to form the Chesapeake Bay Commission (CBC) to coordinate Bay restoration and protection efforts. One of the CBC's first goals was a 40 percent reduction in phosphorus by the year 2000, a goal that it was unable to meet.

However, by 1992 phosphorus had been reduced by 16 percent, mainly as a result of a ban on phosphate-bearing detergents that took effect in January 1989 over the heated objection of detergent manufacturers. In 1992, the average phosphorus content in the Bay was 0.03 mg/L. A representative value for streams that drain urban areas is 0.075 mg/L phosphorus, and those that drain farmland contain about 0.15 mg/L phosphorus.

**Question 21-4:** How much phosphorus was dissolved in the Bay as of 1992? Express your answer in kilograms.

By 2006, despite nearly three decades of study and restoration plans, the Bay was described as an "ecological disaster area" by the Chesapeake Bay Ecological Foundation (CBEF).[7] They reported bacteria levels in the Bay in 2006 to be among the highest measured in any estuarine environment. Zooplankton, the food base for many fish species, were becoming scarce in the Bay's main stem during the summer growing season, in part because comb jellyfish, a major predator of zooplankton and fish larvae, had drastically proliferated.

Potentially dangerous explosions in certain algal populations, called "blooms," are becoming more common in the Bay. Not only are these blooms toxic to many forms of marine life, but they also lower the Bay's already low dissolved oxygen (when they die and decompose) and block sunlight, further restricting the growth of underwater grasses. For example, the Chesapeake Bay Foundation (CBF) reports that

> a 2008 study by the U.S. Geological Survey looked at *Mycroscystis [algal]* blooms around the Bay and found that almost a third of the blooms contained toxins in levels sufficient to make the water unsafe for children to swim in. . . . Due to increased research, since 1996 the number of harmful algal species identified in the Bay has grown from 12 to 34.[8]

The CBF identifies sewage, emissions from coal-fired power plants and motor vehicles, agricultural runoff from farms and concentrated poultry and hog operations, growth in impervious surfaces (roads, parking lots, etc.), and silt from sprawl development as among the reasons for the Bay's decline. By 2011, impervious surfaces covered 1.1 million acres of the Bay's watershed, and were expanding faster than population growth.

---

[6] Cooper, S.R., & G.S. Brush. 1991. Long-term history of Chesapeake Bay anoxia. *Science, 254:* 992–996.

[7] Chesapeake Bay Ecological Foundation, www.chesbay.org.

[8] Chesapeake Bay Foundation, www.cbf.org.

**Question 21-5:**   Based on the area of the watershed we gave you earlier, what percentage of the entire watershed was impervious surface as of 2011?

## THE IMPACT OF SEWAGE ON CHESAPEAKE BAY

The CBF estimated the sewage flow from all sources in the Chesapeake Bay's watershed to be 1.5 billion gallons a day in 2003. Improperly treated sewage is a major source of nitrogen and phosphorus.

**Question 21-6:**   Based on this rate of sewage flow, how long would it take for Chesapeake Bay to fill up with treated sewage, assuming the sewage wasn't flushed to the ocean?

**Question 21-7:**   A 1995 study of the Cameron Run Watershed in Fairfax County, Virginia, found that the amount of phosphorus discharged by urban streams had been significantly underestimated. Researchers found the phosphorus content in these streams to be 2.5 times greater than land-use models had assumed. Discuss the implications for Bay protection in the light of the rapid urbanization of the region.[9]

## AIR POLLUTION AND CHESAPEAKE BAY

Between 20 percent and 35 percent of nitrogen in Chesapeake Bay is from air pollution. A third of this air pollution is from cars, power plants, and farm fields in the watershed itself, but as much as two-thirds is from power plant emissions in Ohio, Kentucky, Michigan, and other states as far away as Alabama[10] (see Figure 21-2). This is apparently due to three factors:

1. Prevailing winds from the southwest, northwest, and west,
2. The lack of control of nitrogen pollution emissions on the part of midwestern coal-burning utilities,

[9] Fairfax County, VA: Cameron Run Watershed Mgmt. Plan, http://www.fairfaxcounty.gov/dpwes/watersheds/cameronrun_docs.htm.
[10] EPA orders twenty-two states to reduce nitrogen oxide emissions. *Washington Post,* October 11, 1997.

3. Smokestacks from the plants that spew pollution high enough into the atmosphere to be carried over 800 kilometers before settling out.

To address water and air pollution problems in the Bay's watershed and throughout the eastern United States, the EPA ordered twenty-two states in an arc from Massachusetts to Missouri to reduce pollutants, including nitrogen (in the form of oxides of nitrogen), or lose federal highway funds.[11] The states most heavily affected and the percentage of nitrogen oxide reductions required are West Virginia (44%), Ohio (43%), Missouri (43%), Indiana (42%), Kentucky (40%), Illinois (38%), Alabama (36%), Wisconsin (35%), Tennessee (35%), and Georgia (35%).

**Question 21-8:**   To what extent should the people in these states be held accountable for air pollution carried beyond their boundaries? Include your reasons and explain them.

These nitrogen oxides contribute to acid precipitation. For example by 2006 parts of the Chesapeake Bay watershed had the most acidic rainfall of any area in the United States, according to the U.S. Geological Survey. Air pollution from excessive nitrogen oxides has been implicated in respiratory problems, especially in children and elderly people. Many allergists and environmental scientists say our health-care system is unfairly burdened by costs to treat persons affected by air pollution from coal-burning power plants and motor vehicles. (You will find a similar observation dealing with turfgrass in Issue 14.) Some believe these costs should be paid in the form of higher electric rates by those who choose coal-fired power plants and who decline to control emissions from these plants.

**Question 21-9:**   Do you agree that the extra costs imposed by using coal to generate electricity should be borne by the consumers of the power? Explain your reasons.

## PFIESTERIA AND MANURE

Some of the largest industrial-scale poultry raising facilities in North America are found within the Chesapeake Bay watershed. On Maryland's Eastern Shore, poultry outnumber people 1000 to 1. And poultry produce 150% more solid waste than humans, per unit area. Moreover, poultry produce 24 times more waste than hogs. An outbreak of a mysterious alga of the genus *Pfiesteria* killed tens of thousands of fish in tributaries of Chesapeake Bay in 1997. The outbreak seemed to be limited to streams that drained areas of Maryland with industrial-sized chicken farms. These farms spread up to 800,000 tons of nutrient-enriched chicken manure each year on fields, some immediately adjacent to streams, far in excess of the nutrients the soil can absorb. The federal government and Maryland proposed

---

[11] Ibid.

to stop the leaching of nutrients from the manure into the streams by paying the chicken growers to leave unplowed grass and, preferably, vegetated buffer strips along streams that drained through their land. The plants would absorb some of the nutrients in runoff, preventing them from contaminating the Bay. The initial cost was reported to be around $250 million.[12]

**Question 21-10:**   To what extent does this $250 million represent a subsidy for chicken production and consumption? Explain your reasoning.

**Question 21-11:**   To what extent, or under what circumstances, do you think government officials have the right to tell farmers and growers what to do with their own land? This involves a key constitutional issue. You can find out more about it by searching the Internet using the phrase "takings clause."

## CHESAPEAKE 2010 AND BAY REMEDIATION

The year 2010 saw "winds of change" in the Chesapeake Bay Watershed, according to the Chesapeake Bay Foundation.

As of 2010:

■ A measure of Bay health reached 45% of that estimated for the Bay in the 1600s.
■ After decades of decline, the Bay's iconic blue crab population "surged 60%."
■ Settlement of a lawsuit required EPA to put the states of the watershed on a strict "pollution diet."
■ Phosphorus loading decreased from 25 million pounds (1985) to 20 million pounds, with a goal of 15 million pounds by 2025.
■ Sediment load decreased from 11 billion pounds (1985) to eight billion pounds.
■ Nitrogen load decreased from 325 million pounds (1985) to 250 million pounds.
■ A "Manure to Energy Summit" was held in 2011 to support plans to turn manure into energy.
■ However, two-thirds of Bay waters failed to meet minimum oxygen standards in 2011, so the health of the Bay declined from "C–" to "D+", mainly due to excessive heat and the impact of intense storms.[13]

---

[12]  MD Dept Natural Res., http://www.dnr.state.md.us/bay/cblife/algae/dino/pfiesteria/pubs.html.
[13]  For details go to www.chesapeakebay.net/pubs/snapc2k.pdf.

### Future Challenges

The Chesapeake Bay Program identified numerous challenges to Bay protection that must be addressed in the decades ahead. Among them are

- Invasive species, brought into the Bay in the ballast water of ocean-going cargo vessels (see Issue 22)
- Recovery of the Bay's oyster population, which had virtually disappeared by 1991
- Protection of forest and agricultural land from sprawl development (see Issues 24 and 25)
- Further removal of nitrogen and phosphorus from sewage (upgrading of all sewage plants in the Watershed could remove 20% of the N from Bay waters)
- Opening thousands of miles of tributaries to fish passage by breaching dams or building fish ladders
- Protecting the blue crab, a symbol of Chesapeake Bay threatened by overharvesting and pollution
- Dealing with material dredged from harbors (Baltimore especially) and shipping channels, which may be contaminated with toxic materials, but which, even if uncontaminated, could smother bottom-dwelling organisms

Here is a summary of the Bay's status from the president of the Chesapeake Bay Foundation, William C. Baker: "Despite progress on many fronts, the Bay's health has stalled at a dangerously low level."

**Question 21-12:** Explain why a healthy Chesapeake Bay is an important component of sustainability for the Chesapeake Bay watershed.

**Question 21-13:** Summarize the major points of this Issue.

## FOR FURTHER THOUGHT

**Question 21-14:** In addition to sewage, sources of phosphorus in Chesapeake Bay include sediment runoff, agricultural runoff, and animal waste from vast chicken and pig farms. Research and discuss any or all of the following questions.

1. How does sediment contribute phosphorus?
2. What is the effect of construction activity on sediment runoff and thus on phosphorus?
3. Why is agricultural runoff other than manure a major source of phosphorus?
4. How might phosphorus from sediment and agricultural runoff be minimized?
5. What can be done about animal waste?

**Question 21-15:**   Go to www.cbf.org and assess the status of Bay remediation for yourself. Also see www.chesbay.org/acidRain to assess the impact of acid rain (much of it from outside the watershed) on the Bay's ecosystem, especially fish.

**Question 21-16:**   Research the movement of water out of the Bay by searching the Web using the term "estuarine circulation." How efficiently does the freshwater mix with the saltwater in Chesapeake Bay? Does the seawater flush pollutants out of the Bay? If so, where do they go?

**Question 21-17:**   Scientists studying satellite imagery have concluded that a significant amount of phosphorus was being introduced into Chesapeake Bay from sewage treatment plants in Pennsylvania, New Jersey, and New York. Study marine circulation in the coastal North Atlantic (in an oceanography text, for example) and form a hypothesis to explain how this could happen.

**Question 21-18:**   Discuss the general impact of population growth (leading to increased sewage) on the health of Chesapeake Bay.

**Question 21-19:**   The Chesapeake Bay watershed has been hit by a number of damaging hurricanes since European colonization. Go to http://chesapeake.usgs.gov/, the USGS's Chesapeake Bay activities site. Then assess the likelihood of flood damage from hurricanes of increasing severity from ocean warming based on the region's population growth, high property values, and soil permeability.

**Question 21-20:**   There are, of course, other threatened estuaries in the United States. In the west, restoration of San Francisco Bay is the subject of a concerted effort on the part of the state of California, the U.S. Bureau of Reclamation, agricultural interests in the Central Valley, and a number of public interest and environmental groups. Research the issues involved. Include in your study how water is diverted from the river systems that feed the San Francisco Bay in order to grow crops in the arid Central Valley and to water lawns.

# ILLEGAL IMMIGRATION:
## *BALLAST WATER AND EXOTIC SPECIES*

## KEY QUESTIONS

- What are exotic species?
- How are they able to cross oceans?
- What traits allow them to colonize foreign environments?
- What threats do they pose to ecosystems?
- What can we do to protect water bodies from nonnative species?

## INTRODUCTION

In September 1999, then U.S. President Bill Clinton issued an executive order that directed the Departments of Agriculture, Interior, and Commerce, the EPA, and the U.S. Coast Guard to develop an Alien Species Management Plan to deal with the economic, ecological, and health impacts of *invasive species*. As a result, the National Invasive Species Council (NISC) was set up, with its own web site.[1] The most recent iteration of the Invasive Species Management Plan was prepared in 2008 and covered the period 2008–2012. It set up a five-point approach to invasive species management. Foremost was *prevention*—keeping invasive species out in the first place. Next was *early detection and rapid assessment*, leading to *rapid response. Control and management* techniques were applied to established species, and finally *restoration* of affected habitats after species had been eradicated. Unfortunately, no invasive species, once established, has ever been eradicated.

But, what are invasive species anyway?

## ILLEGAL IMMIGRANTS

Silently, almost imperceptibly, the planet's waters and land are being invaded by plants, animals, bacteria, and even viruses from distant climes. These organisms are called alien, exotic, or invasive species. Sometimes their impact is negligible, rarely is it beneficial, and often it borders on the disastrous. More than a decade ago Cornell University professor David Pimentel estimated the total cost of invasive species at $123 billion a year.[2] The NISC estimates the cost of just six species at $74 billion a year.

Instead of remaining in an ecosystem in which all members have evolved and interacted over time, in the relative blink of an eye invasive species may be transported beyond their

---

[1] National Invasive Species Council, http://www.invasivespecies.gov.

[2] Pimentel, D., L. Lach, R. Zuniga, and D. Morrison. 2000. "Environmental and Economic Costs of Nonindigenous Species in the United States." *Bioscience*, 50(1): 53–56.

natural range into the presence of other organisms with which they will immediately begin to interact and perhaps compete. Once thrust into a new environment, an organism faces a new set of conditions. To survive, all living organisms must live long enough to bear offspring and thus ensure the future of their gene pool. The "aim" of exotic species is not to take over an estuary or clog a factory's water pipes, but to simply survive and reproduce.

Scientists believe that most nonnative organisms fail to survive in their new environment long enough to become established. And that's a very good thing. But occasionally, an introduced organism finds its new home completely livable, sometimes even ideal. Successful invasive species usually share a similar set of characteristics, according to the U.S. Coast Guard:

- They are *hardy*, indicated by their surviving a trip inside a ship for perhaps thousands of miles.
- They are *aggressive*, with the capacity to outcompete native species.
- They are *prolific breeders* and can take quick advantage of any new opportunity.
- They *disperse rapidly*.

A planktonic larval stage facilitates rapid dispersal of aquatic species, which allows the juveniles to be carried far and wide by currents. Such an introduced species often spreads rapidly, especially when predators and pathogens normally encountered in its home range are absent from the new environment or when they are better able to feed than their new neighbors. Or if they find their new neighbors especially tasty!

In the above scenario, alien species flourish and potentially can reach astonishingly high population levels. Often, native species are displaced by the invaders. Then the situation is often called an *invasion*.

Invasive species can damage an ecosystem by

- Out-competing native species
- Introducing parasites and/or diseases
- Preying on native species
- Adversely altering habitat[3]

# BALLAST WATER

*Ballast water* is carried by ships in special tanks to provide stability and optimize steering and propulsion. Most invasive species are brought to new shores in the ballast water of ships (Figure 22-1), or attached to ship's hulls (but animals dumped into an estuary from aquariums or accidental releases from aquaculture facilities may also contribute). Ballast water was one of the major pathways for exotic aquatic organisms such as the Chinese mitten crab[4] (annual economic cost unknown), green crab (annual economic cost $44 million), and Asian clam (annual economic cost $1 billion), which threaten native marine life in San Francisco Bay and as far north as Washington state.

The West Coast invaders are driving out native crabs and clams and threatening local oysters—even burrowing into and weakening flood control levees, which could potentially

---

[3] San Francisco Estuary Institute, http://www.sfei.org/node/2210.

[4] The U.S. Bureau of Reclamation operates a series of giant pumps at Tracy, California, on the delta of the Sacramento River to ship San Francisco Bay water to southern California. During 1998, they found so many mitten crabs clogging fish screens (which keep fish out of the pumps and so keep them from being cut to pieces) that the agency spent $400,000 to build a series of "crab screens" that catch the crabs before they clog the fish screens. The devices fling the crabs onto a conveyor belt for removal by a firm that pays for the privilege. The firm uses the mitten crabs for bait. Ironically, Chinese mitten crabs are a treasured delicacy in Hong Kong, and some entrepreneurs have explored shipping the California crabs to Hong Kong. But the state refuses to allow this, fearing that it will encourage more importation of the crabs and other exotic species with more unforeseen consequences! (*San Jose Mercury News.* Crab migration drops off, by N. Vogel. Oct 14, 1999.)

**FIGURE 22-1** Ship dumps ballast water after entering harbor. (Photo by L. David Smith. Courtesy of Northeast Sea Grant.)

result in huge losses from property damage during floods.[5] Another such alien species is the veined rapa whelk, which was discovered in Chesapeake Bay in the late 1990s. And perhaps the most infamous example is the zebra mussel, which has infested the Great Lakes and is spreading outwards.

The problem with ballast water is very simply stated. Ballast water is taken up by a ship in ports and other coastal regions whose waters may be rich in planktonic (small, floating, or weakly swimming) organisms. It may be released at sea, in a lake, or a river, or in coastal waters—wherever the ship reaches a new port. As a result, a myriad of organisms is transported around the world within the ballast water of ships and is released.

Since ballast water management involves interstate and international commerce, individual states and counties cannot, under the U.S. Constitution, regulate ballast water. But the federal government can. And the task of ballast water management has been delegated to EPA and the Coast Guard (USCG). The USCG now requires that international vessels dump their ballast water, and take on new water, at least 200 nautical miles from shore in water depths of at least 2,000 meters. Furthermore, vessels must certify that their ballast water does not contain hazardous levels of potentially invasive species. And ballast water must meet standards proposed by the International Maritime Organization for maximum organisms per unit volume of ballast water. These new rules will affect as many as 20,000 domestic and international commercial vessels a year.[6] Similar proposals are being considered by international agencies. There are about 50,000 large commercial vessels sailing the oceans.

Before the new rules, the National Oceanographic and Atmospheric Administration (NOAA) calculated that 40,000 gallons (150,000 L) of foreign ballast water were dumped into U.S. harbors each *minute*.

Here are two examples of the impact of ballast water:[7]

■ Scientists studying an Oregon bay counted 367 types of organisms released from ballast water of ships arriving from Japan over a four-hour period!
■ Another study documented a total of 103 aquatic species introduced to or within the United States by ballast water and/or other mechanisms, including seventy-four foreign species.

---

[5] California Dept of Fish and Game, http://www.dfg.ca.gov/delta/mittencrab/life_hist.asp.
[6] U.S. Coast Guard, http://www.uscg.mil/hq/cg5/cg522/cg5224/bwm.asp.
[7] Ships' Ballast Water and the Introduction of Exotic Organisms in to the San Francisco Estuary—Current Status of the Problem and Options for Management. *California Urban Water Agencies,* October 1998.

Today, ballast water appears to be the most important means by which marine species are transferred throughout the world. To the extent these unwelcome visitors do economic damage, they make up a generally hidden cost of world trade.

## INVASIVE SPECIES ALONG THE PACIFIC COAST

Chinese mitten crabs, another invasive species affecting San Francisco Bay and the Sacramento River Delta (and, apparently now, Chesapeake Bay), are described by some scientists as "burrowing fiends," digging burrows that can significantly weaken levees (embankments built to prevent flooding) in a region that is prone to dangerous floods.

**Question 22-1:** If the costs incurred during a flood are in part due to ships involved in international trade, how can these costs be fairly apportioned? Is it fair for only those people who are affected by floods in California to pay for the hidden costs incurred as a result of invasive species? Is it reasonable to price imported goods cheaply and then expect local residents to bear the cost of flooding resulting from this trade? Suggest some possible solutions to this problem, but remember, localities and states do not have the right under the Constitution to regulate international trade.

Another threat to the Pacific Coast is posed by the European green crab. This crab is a recent import to coastal California, and it seems to be migrating up the coast to Oregon and Washington and beyond. An aggressive predator, it prefers clams to oysters, but it could prey on baby Dungeness crabs (an economically important species) and smaller shore crabs. Washington oyster farmers are already uneasily coexisting with another exotic: the oyster drill that came from Japan with Pacific oysters.[8] Washington oyster growers have already had to abandon habitat overrun by the oyster drill.

By now you have seen how invasive species can materially affect a region's economy as well as its environment.

## INVASIVE SPECIES AND CHESAPEAKE BAY

There is growing concern about invasive species' impact on Chesapeake Bay (see Issue 21). Consider these 1995 statistics from the Chesapeake Bay Commission:[9]

■ More than 90 percent of vessels arriving at Chesapeake Bay ports carried live organisms in ballast water, including, but not limited to, barnacles, clams, mussels, copepods, diatoms, and juvenile fish.

■ Nonindigenous species have been responsible for paralytic shellfish poisoning, declining commercial and sport fisheries, and even cholera outbreaks!

■ The ports of Baltimore and Norfolk received 2,834,000 and 9,325,000 metric tonnes of ballast water, respectively, in 1995. This water originated in nearly fifty different foreign ports.

In 2002, a database of organisms not indigenous to Chesapeake Bay prepared by the Marine Invasions Research Lab of the Smithsonian Environmental Research Center lists

---

[8] Washington Sea Grant, http://www.wsg.washington.edu/.
[9] Chesapeake Bay Commission, http://www.chesbay.us/.

more than 160 species and classifies another forty-two as of uncertain origin.[10] In 2006, a single specimen (an adult male) of a Chinese mitten crab was captured in the Bay.

### The Veined Rapa Whelk (*Rapana venosa*)

*Rapana venosa* (Figure 22-2) is a predatory gastropod. Juliana Harding and Roger Mann, two researchers at the Virginia Institute of Marine Sciences (VIMS), are studying the whelk's impact on Chesapeake Bay. The following is a summary of some of their research.[11]

Discovery of *R. venosa* was purely by accident: A routine trawl in the lower reaches of Chesapeake Bay turned up an unknown organism, which was ultimately identified by scientists at the Smithsonian Institution in Washington, D.C., and by a Russian biologist at the Moscow Academy of Sciences as *R. venosa. R. venosa* has left a trail of destruction behind in its wanderings, including decimating an oyster population in the Black Sea.

The researchers conducted another trawl, which yielded two live masses of *R. venosa* eggs, which they returned to the lab and set about to hatch. The scientists were eager to determine the tolerance of the hatchlings to variations in temperature and water salinity.

## Have You Seen This Animal?
### The Veined Rapa Whelk
### (*Rapana venosa*)

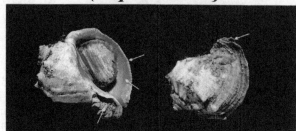

The Virginia Institute of Marine Science (VIMS) is interested in any sightings of this large snail in Virginia waters. The veined rapa whelk is native to the Sea of Japan, reaching sizes of 5 to 7 inches in length. There are several distinguishing characteristics that are highlighted by arrows in the above pictures. Note the small teeth along the edge of the shell and the orange coloration along the inner edge of the shell. Other characteristic features are a pronounced channel (columella) and the ribbing at the lower end of the shell.

**FIGURE 22-2** The veined rapa whelk. By 2009, more than 19,000 whelks had been captured in Chesapeake Bay as part of a bounty program. U.S. East Coast estuaries have favorable temperatures and ample prey (bivalves) for the rapa whelk. ((Juliana Harding, Molluscan Ecology Program, Virginia Institute of Marine Science. Courtesy of Virginia Institute of Marine Science.)

---

[10] Smithsonian Institution, http://www.serc.si.edu/labs/marine_invasions/.

[11] Harding, Juliana and Roger Mann, 2002, Molluscan Ecology Program, Dept. of Fisheries Science, VIMS, Gloucester VA; Mann, R., J. M. Harding, & E. Westcott. 2006. Occurrence of imposex and seasonal patterns of gametogenesis in the invading veined rapa whelk *Rapana venosa* from Chesapeake Bay, USA. *Mar. Ecol. Prog Ser.* 310: 129–138.

**Question 22-2:**   Why do you think they were interested in these data?

The scope of potential contamination of the Bay by *R. venosa,* which is native to the Sea of Japan, and other introduced species can be appreciated by the 15 million tonnes of ballast water dumped into Chesapeake Bay ports during 1998 solely by ships from ports with active *R. venosa* populations. And, since the entrance to Chesapeake Bay is the site of considerable coastal shipping, infestations of harbors from Boston to Charleston, South Carolina, remains a possibility.

Subsequently, Harding and Mann learned that whelks preferred hard, bottoms into which they would quickly burrow. A 6-inch whelk could completely hide itself in less than an hour, leaving only its purplish siphon exposed, which it would instantly withdraw if disturbed. *R. venosa* spends at least 95 percent of its life burrowed, but it can and does move while burrowed, at speeds up to one body length per minute. They learned that the whelk can feed and mate while completely buried.

**Question 22-3:**   Considering the whelk's preferred habitat, propose how scientists could determine its potential range in an estuary like Chesapeake Bay.

*R. venosa*'s preferred diet is hard clams, but it can eat oysters, soft clams, or mussels if its favorite food is unavailable. Unfortunately for the clams, they share the whelk's habitat. And finally, there is a "healthy" hard clam commercial fishery in Chesapeake Bay.

The researchers were also interested in the whelk's predators, if any, in *R. venosa*'s home waters. There are few: Octopi eat the whelks in the waters of southern Russia and the Black Sea, but there are no octopi in Chesapeake Bay. Other native whelks in Chesapeake Bay prey on smaller individuals of *R. venosa,* but there is an interesting twist: *R. venosa*'s shell is much thicker than that of native whelks. Moreover, *R. venosa*'s boxy shape means the creature is hard for other whelks to eat as adults. So if the whelks can survive to adulthood, they have little to fear from the natives in Chesapeake Bay.

The researchers next turned their attention to what could prove to be *R. venosa*'s "Achilles' heel," the egg and juvenile stages. They concluded that the whelks may be most vulnerable as eggs. Migrating fish could eat the bright yellow egg cases or dislodge them, causing damage and perhaps death to the developing eggs. Recent findings indicate that early life stages are hardy and that juveniles are generalists in eating and are tolerant of a wide range of salinities.

Mann and Harding's research has shown that this species exhibits broad environmental tolerance, uses a reproductive strategy in which they mature at a small size and young age, live about fifteen years, have a high fecundity, and lack predators and competitors in their new habitat—a recipe for disaster.[12]

Figure 22-3 shows the distribution of the whelk as of August 2009 in Chesapeake Bay.

---

[12] Ibid

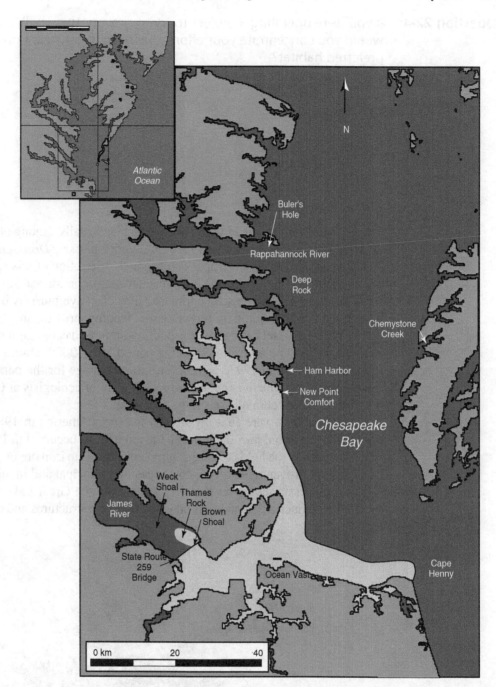

**FIGURE 22-3** The distribution of veined rapa whelks (lightest shading) in Chesapeake Bay, 2009. (After Harding and Mann, 2005, *Journal of Shellfish Research*, vol. 24(2): 381–385, and http://web.vims.edu/mollusc/research/rapaw/merapmap.htm)

**Question 22-4:** If you were operating a project to try to prevent the whelk's further spread, where would you concentrate your efforts, based on the whelk's present distribution and preferred habitat?

## The Zebra Mussel

The phylum Mollusca includes clams, oysters, and snails, among others. The most notable invasive mussel introduction so far is the zebra mussel, *Dreissena polymorpha* (Figure 22-4), a native of eastern Europe. The original description of this species, from 1769, was of populations in the Ural River and Caspian Sea of the former Soviet Union.

Zebra mussels can destroy entire colonies of native mussels by interfering with such basic functions as respiration, reproduction, feeding, growth, and movement.

The zebra mussel has caused serious economic and ecosystem impacts, with costs projected in 1998 to be $5 billion over the period 1998–2008, absent controls. But a similar estimate published in 1993 forecast $3 billion damage for the period 1993–2003, so you can see that the problem is getting worse. A team of ecologists at Cornell University estimated the cost of zebra mussels to be $3 billion a year.[13]

Zebra mussels were first discovered in North America in 1988 in Lake St. Clair in Michigan. The initial introduction is believed to have occurred in 1985 or 1986 via ballast water, perhaps in the holds of cargo ships sent to pick up iron ore or grain. Since then, zebra mussels have proven to be a very costly pest to municipal and industrial water users.

The impact on industries drawing water from the Great Lakes was rapid and caused severe flow reductions as mussels attached to intake structures and the insides of pipelines.

**FIGURE 22-4** Zebra mussels cover a crayfish. This invader has spread to many U.S. rivers and lakes and can tolerate estuarine water. (Courtesy of Ontario Ministry of Natural Resources.)

---

[13] www.glc.org/ans/96rpt.html.

Utilities that operate power plants relying on lake water for cooling are among the most heavily impacted. Since 1989, power plants, water utilities, industrial facilities, and navigation lock and dam operators have spent more than $70 million trying to control and manage zebra mussel infestations.[14]

One of the reasons for zebra mussel success is their proficiency at breeding. Up to 1 million eggs can be laid by one female in a spawning season. Upon hatching, the larvae may be dispersed by currents. As juveniles, they settle to the bottom and attach. Importantly, they have trouble keeping attached in water velocities above around 2 meters per second.

Moreover, zebra mussels are filter feeders, which means they have specialized organs for filtering food, mainly algae, out of the water. Fortunately, many organisms in the Great Lakes feed on zebra mussels.

**Question 22-5:**   Suggest a way in which the introduction of zebra mussels into a water body contaminated with too many planktonic algae might actually be beneficial.

**Question 22-6:**   In the Great Lakes, zebra mussel concentrations of up to one million per square meter have been reported by the USGS. At this density, how many zebra mussels were there per 1 square centimeter?

**Question 22-7:**   An average-size zebra mussel filters water at a rate of about 1 gallon of water a day.[15] At the above density, how much water could have been filtered clean of algae by 1 square meter of zebra mussels each day?

**Question 22-8:**   At this density, how many zebra mussels would be needed to filter a cubic kilometer of algal-polluted lake water each day? This would be equal to a lake with an area of 100 square kilometers and an average depth of 10 meters.

---

[14]  Great Lakes Science Center, www.glsc.usgs.gov.
[15]  *Journal of Great Lakes Research*, http://www.glerl.noaa.gov/pubs/fulltext/1995/19950011.pdf.

**Question 22-9:** Assess the potential adverse impact of introducing zebra mussels into a polluted water body to clean it of algae.

Once zebra mussels become established in a water body, it is impossible to eradicate them using any present technology. Figure 22-5 shows the 2011 distribution of zebra mussels.

**Question 22-10:** Based on Figure 22-5, should control of zebra mussels be primarily a local, state, or federal responsibility? Explain.

It is clear that without a massive effort (and perhaps even with one) the spread of zebra mussels will not be contained.

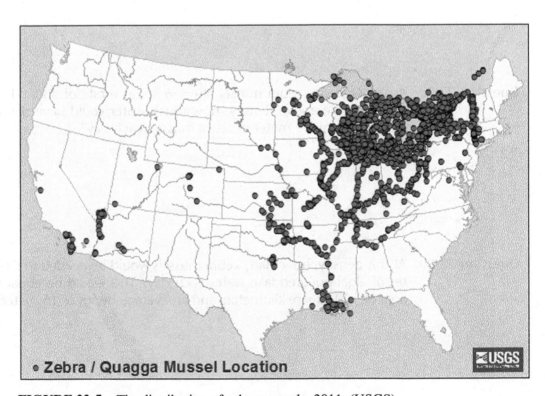

**FIGURE 22-5** The distribution of zebra mussels, 2011. (USGS)

**Question 22-11:** Summarize the main points of this Issue.

**Question 22-12:** Discuss the issue of invasive species from the perspective of sustainability and sustainable communities.

## FOR FURTHER THOUGHT

**Question 22-13:** Go to http://www.npr.org/templates/story/story.php?storyId=154658739, and read about or listen to the report of invasive species from the Japanese tsunami of 2011 washing up onto the Oregon coast. In one or two paragraphs, summarize the report.

# CATCH OF THE DAY:
## *THE STATE OF GLOBAL FISHERIES*

## KEY QUESTIONS

- What is the state of global fisheries?
- What are environmental impacts of commercial fishing?
- How much seafood do we eat?
- How important is seafood as a protein source?
- Do sustainable fisheries exist?
- Is aquaculture sustainable?

## REPORT CITES DIRE CONDITIONS OF EUROPE'S FISHERIES

Fishing should be banned in almost a third of UK waters, according to a Royal Commission on Environmental Pollution report released in December 2004.[1] More than 40 percent of commercial fish species in the northeast Atlantic and adjacent seas are being caught at higher than sustainable rates. Chairman Sir Tom Blundell said, "It is hard to imagine that we would tolerate a similar scale of destruction on land but, because it happens at sea, the damage is largely hidden."[2]

A day after the Royal Commission report was published, the European Union proposed major fishing cuts. The measures include the closure of heavily depleted cod grounds in the North Sea, the Irish Sea, and off the west coast of Scotland. They also included reductions of up to 60 percent in herring catches, 34 percent for cod, and 27 percent for mackerel. Europe's then-fisheries commissioner Joe Borg said the proposals were "a balance between what is biologically necessary and what is economically reasonable."[3]

**Question 23-1:** Assess the concept of striking a "balance between what is biologically necessary and economically reasonable." Who should decide what is "economically reasonable?" What is "biologically necessary?" With which group would you have the most confidence? Why?

---

[1] www.rcep.org.uk/fishreport.htm.
[2] "Huge No-Fishing Zones 'Offer Only Hope' of saving Marine Ecosystem from Disaster" by Michael McCarthy *The Independent* (UK) 12/8/04.
[3] "EU set to close fishing grounds." BBC online http://news.bbc.co.uk 12/8/04.

In the United States, Whole Foods Markets, a nationwide chain of more than 170 grocery stores, announced in 2012 that, in addition to previously banned orange roughy, bluefin tuna and shark, it would no longer sell Atlantic halibut, Atlantic grey sole, octopus, skate wing, sturgeon, most swordfish, trawl-caught Atlantic cod, (certain) tuna, turbot, imported wild caught shrimp, and rockfish, because these species are endangered due to overfishing.[4]

Will the other more than 225,000 U.S. retail food stores follow suit? Are the species in question, and others, really endangered?

## THE IMPORTANCE OF GLOBAL FISHERIES

Even though less than 1 percent of global caloric intake comes from fish, the importance of fisheries to the global and many national economies cannot be overstated. Consider the following information on fisheries:[5]

- The value of the international fish trade for 2004 was $71 billion.
- U.S. fishery product exports amounted to $5.7 billion in 2009.
- Canned tuna sold to U.S. consumers declined from 980 million pounds in 2000 to 763 million pounds in 2009.
- Aquaculture accounted for 46% of fish supplies in 2009.
- The economies of many countries such as Iceland, Peru, and Norway depend heavily on fish product exports.
- In eastern Canada, the closure of the cod fishery cost at least 40,000 jobs in a country with a population one-tenth that of the United States.[6]

Although the contribution of fish to the diet of humans may seem small if we simply count calories, it becomes critical if we consider protein. Sixteen percent of animal protein globally comes from fish. In the Far East, where most people live, a quarter of animal protein comes from fish. In developing countries worldwide, where population growth is ominously high, more than 1 billion people depend on fish as their primary source of protein.[7]

In addition, fish is considered by affluent westerners to be a "health food." Fatty fish like salmon and mackerel have relatively high levels of omega-3 fatty acids, which have been shown in clinical studies to reduce the risk of heart attack by 50 to 70 percent. Finally, fish and fish by-products, representing as much as one-third of wild-caught fish, are a mainstay of the pet food industry and are used as a constituent of animal feed as well.

## ARE WE RUNNING OUT OF FISH?

All-you-can-eat seafood restaurants and seafood shops abound. World per capita consumption was 17.2 kilograms in 2009.[8] Are there enough fish to sustain this level of use? To sustain an ever-growing demand?

Humans obtain fish and shellfish in three ways: marine capture fisheries, fresh water capture fisheries, and aquaculture, or "fish farms." Global forecasts of upper limits to capture fisheries have been made since the early 1970s. According to the Food and Agriculture Organization (FAO) of the UN, ". . . about half of the main stocks or species groups are fully exploited and are therefore producing catches that have reached, or are very close to, their maximum sustainable limits." These stocks, in other words, offer no reasonable expecta-

---

[4] www.wholefoods.com.
[5] Statistical Abstracts of the U.S., UN FAO
[6] http://www.cbc.ca/news/background/fishing/cod.html.
[7] Food and Agricultural Organization of the United Nations. 1993. Marine fisheries and the law of the sea: A decade of change. FAO Fisheries Circular No. 853. Rome; World Resources Institute. 1996. *World resources 1996–97: A guide to the global environment.* (New York: Oxford University Press); www.fao.org/.
[8] www.fao.org.

tions for further expansion. Another quarter of stocks or species groups are overexploited, or recovering, which means these fisheries could crash and disappear as a food source, like Atlantic cod. Globally, total fisheries production has plateaued.

Chinese production has continued to increase while global production outside China has not. The top capture fisheries nations are China, Peru, the United States, Indonesia, Japan, Chile, and India. According to the Worldwatch Institute, industrial fleets fished out at least 90 percent of all large ocean predatory fish—tuna, marlin, swordfish, sharks, cod, halibut, skates, and flounder—in just fifty years.[9]

## ENVIRONMENTAL COSTS OF FISHING

Commercial fishing can be a destructive activity. In addition to endangering the target species, many fishing methods destroy habitat. Consider trawling, typically done for shrimp and other bottom-associated species like Atlantic cod. In this method, a 10- to 130-meter (33–426 ft) net is scraped across large areas of the bottom, collecting virtually everything in its path, including endangered sea turtles. Trawling, which became popular with the advent of the diesel engine in the 1920s, is practiced worldwide on virtually every different bottom type.[10]

According to the U.S. National Academy of Sciences:[11]

■ Repeated trawling and dredging result in harmful changes in benthic communities, some likely permanent.

■ Bottom trawling reduces the productivity of benthic habitats.

■ Fauna that live in low natural disturbance regimes are generally more vulnerable to disturbance.

Saturation trawling, in which the net is repeatedly fished in an area until virtually no fish or shrimp are left, has been compared to clear-cutting a forest. The comparison is appropriate: Trawling heavily damages sessile benthic organisms like sponges, hydroids, and tube-dwelling worms and displaces associated fauna like fish and crustaceans. Complete recovery in both cases may take centuries. The comparison between trawling and forest clear-cutting breaks down when one considers the area involved. Approximately 100,000 km$^2$ (38,000 mi$^2$) of forest are clear-cut annually, whereas an area 150 times as large is trawled. Extremely destructive methods like dynamiting and poisoning are still used in some areas.

A large percentage of fish caught is considered *bycatch* (Figure 23-1); that is, juveniles too small to eat or nonfood species. Bycatch is also known as *bykill* because, while some bycatch is returned to the water and may survive being caught, most die in nets or on longlines or are returned to the water dead or dying. As much as 90 percent of the trawl contents may be non-target and hence unused species, sometimes called *trash fish* by fishers. As many as 25,000 fishing boats ply U.S. waters.

The U.S. imports wild-caught shrimp from nearly forty countries. As you saw above, Whole Foods Markets no longer sell foreign wild-caught shrimp. Although shrimp trawling may cause extreme environmental damage, most consumers are unaware of the dimensions of its destructiveness. Bycatch also includes animals such as sea turtles, dolphins, seals, and whales trapped accidentally in fishing gear. Seabirds, including endangered species like albatrosses, can drown when they take baited hooks and are pulled under water.

---

[9] www.worldwatch.org.

[10] Waitling, L., & E.A. Norse. Disturbance of the seabed by mobile fishing gear: A comparison to forest clearcutting. *Conservation Biology, 12:* 1180–97.

[11] National Academy of Sciences, www.nap.edu/books/0309083400/html/; http://books.nap.edu/books/0309083400/html/20.html#pagetop.

**FIGURE 23-1** Bycatch/bykill: A large sunfish trapped in a net and drowned. (Courtesy of Roger Grace/Greenpeace.)

**Question 23-2:**  Assess the use of the term "trash fish" for living organisms. Loggers call trees they can't sell "trash trees." What is your response to these terms? Why do you think fishers and loggers use them?

Habitat loss, including mangrove destruction from shrimp farming, is one impact of aquaculture (see "For Further Thought" at the end of this Issue). In Ecuador alone over 200,000 hectares (approximately 500,000 acres) of mangrove ecosystems and adjacent landscape have been converted to shrimp aquaculture.[12]

Finally, the environmental cost of commercial fishing extends to the pollution arising from the use of equipment (like fishing boats) and supplies; fuel spills; and transportation and refrigeration of fishery products.

## OF SLIMEHEADS AND PATAGONIAN TOOTHFISHES

Would you eat a fish called a slimehead? Or a Patagonian toothfish? Probably not, so marketing specialists renamed them "orange roughy" (Figure 23-2a) and "Chilean sea bass" (Figure 23-2b). The former was popular in the mid- to late 1980s, whereas the latter's popularity peaked in the early 2000s. Both are still available at many fish markets and seafood restaurants.

Unfortunately, renaming the fish changed neither their biology nor their fate. The orange roughy is a classic example of failing to gain a complete understanding of a species' biology before exploiting it as a fishery. The Chilean sea bass is yet another reminder that we refuse to learn from our mistakes. Both cases demonstrate the power and potential environmental destructiveness of effective marketing.

---

[12]  McGinn, A.P. 1999. Safeguarding the health of the oceans. *Worldwatch Paper* 145, Worldwatch Institute.

A

World Wide Web ver., v. 01-09-98
RFE Team: Tenge, Barnett, Savary, Rogers, Dang
RFE Funding: OS/CFSAN and ORA
RFE contact: btenge@fdaem.ssw.dhhs.gov
RFE WWW coord.: F. Fry (CFSAN)
Internet: frf@fdacf.ssw.dhhs.gov

Species: *Dissostichus eleginoides* Smitt, 1898
Common Name: AFS-Patagonian Toothfish
RFE Code (this specimen): disseleg I94-001
Photographer: W. Savary (SAN-DO)
Date: 09-12-94
Image #: 287

Scanner: DTS-103AI Drum
Filename: pnrd001.tif
Date: 08-09-95
Original File: 12Mb, 400dpi
Orig. Image arch.: SEA-DO
Tissue arch.: SEA-DO/SAN-DO
Fish Provided by: USDC
No Authentication

B

**FIGURE 23-2** (a) The erstwhile *slimehead*, successfully marketed as the orange roughy. (Courtesy of JACANA/Photo Researchers, Inc.) (b) The fish formerly known as the Patagonian toothfish, now available in Western restaurants and seafood shops as the Chilean sea bass. It is not at all closely related to the group of fishes commonly known as sea basses. (W. Savary, Center for Food Safety and Applied Nutrition).

The orange roughy fishery began off New Zealand in 1978 and quickly exceeded 35,000 tons ($31.8 \times 10^6$ kg).[13] Unfortunately, because the species is long-lived (adults may live to be 150!)), reaches sexual maturity late in life, and doesn't produce profuse numbers of offspring, the harvest had by 1990 been reduced over 70 percent. Maximum sustainable yield in Australia, the amount fisheries managers estimate can be annually harvested without damaging the population, is 665 tonnes until the stocks are rebuilt, then 995 tonnes. The 2004 Australian harvest was 1600 tonnes. [14]

Environmental Defense recently issued an alert due to high levels of mercury found in some commercial orange roughy, and recommends substituting U.S. catfish or farmed sea bass.[15]

[13] www.h-economica.uab.es/wps/2001_04.pdf.

[14] www.h-economica.uab.es/wps/2001_04.pdf.

[15] Monterey Bay aquarium, at http://www.montereybayaquarium.org.

Chilean sea bass have a similar biology and are caught in deep water in Antarctic seas. While the U.S. Department of Commerce denies the species is overfished[16] many environmental groups vehemently disagree, especially considering that many are illegally caught. Environmental Defense, for example, says the species is "severely overfished." The U.S. imports about 10,000 tonnes of Chilean sea bass a year, between 15-20% of the world catch.

## ANALYZING THE WORLD FISH CATCH

Data on the total annual world fish catch are compiled by the FAO. Table 23-1 contains this information (including aquaculture for 2004-2009) for the years 1950 to 2009.

**TABLE 23-1** ■ Global Wild Fish Catch, Aquaculture Production, and Population, 2004–2009

| Year | Catch (Mt*) | Aquaculture (Mt) | Global Population (Bil) |
|------|-------------|------------------|--------------------------|
| 1950 | 19 | — | 2.56 |
| 1960 | 36 | — | 3.04 |
| 1970 | 58 | — | 3.71 |
| 1980 | 67 | — | 4.45 |
| 1990 | 86 | — | 5.28 |
| 2000 | 96 | — | 6.16 |
| 2004 | 95 | 41.9 | 6.38 |
| 2005 | 92.3 | 44.3 | 6.45 |
| 2006 | 89.8 | 47.3 | 6.6 |
| 2007 | 89.9 | 49.9 | 6.7 |
| 2008 | 89.7 | 52.6 | 6.79 |
| 2009 | 90.0 | 55.1 | 6.84 |

\* Million Tonnes
Source: FAO, *Yearbook of Fishery Statistics: Capture Production and Aquaculture Production*; population data from U.S. Department of the Census.

**Question 23-3:** On the axes below, make a graph of world fish catch (in million tonnes) for the period from 1950 to 2004. Interpret the graph: What trends can you infer?

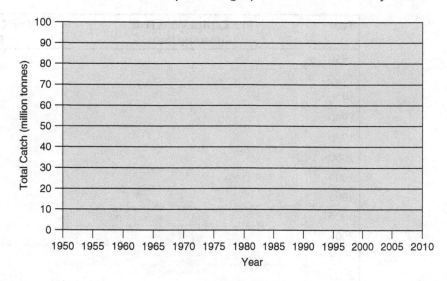

**Question 23-6:** Calculate the average annual rate at which total catch changed for the ten-year periods 1950–60, 1960–70, 1970–80, 1980–90,1990–2000, and 2000–2009. Use the compound growth equation: $r = (1/t) \ln(N/N_0)$, where r = the percentage increase per unit time; t = the number of years; N = total catch at the end of the given time; and $N_0$ = total catch at the beginning of the period (see "Using Math in Environmental Issues," pages 6–8.)

**Question 23-5:** On the same graph above, plot aquaculture production from 2004–2009. Calculate the average annual rate of change over that period.

**Question 23-6:** A July 13, 2003, press release from the U.S. House of Representatives Committee on Resources stated ". . . we know, contrary to the conclusions of the recent Pew Oceans Commission report, that there are indeed still plenty of fish in the sea." Is this statement consistent with the trends you just observed? Explain your reasoning. Discuss the future of capture (i.e., nonaquaculture) fisheries.

**Question 23-7:** Use the data in Table 23-1 to calculate per capita (i.e., per person) wild fish consumption. Report as kg/person (i.e., "kilograms per person"). Fill in the empty column below with your calculations.

| Year | Per Capita Wild Fish Consumption | Per Capita Aquaculture Consumption* |
|---|---|---|
| 1950 | | — |
| 1960 | | — |
| 1970 | | — |
| 1980 | | — |
| 1990 | | — |
| 2000 | | — |
| 2004 | | — |
| 2005 | | |
| 2006 | | |
| 2007 | | |
| 2008 | | |
| 2009 | | |

* 2004–2009

**Question 23-8:**   On the axes below, make a graph of wild per capita fish consumption (in kg/person) for the period 1950–2009, and aquaculture per capita consumption 2004–2009. Describe any trends in your graph.

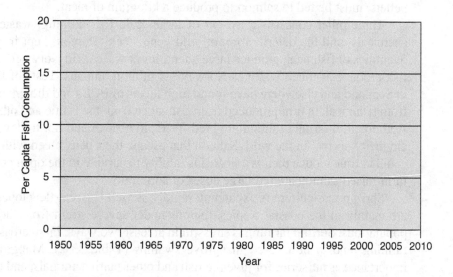

**Question 23-9:**   With global population growing at about 1.1 percent per year, do you think that capture fisheries alone can supply demand on fish stocks of a burgeoning human population? Justify your answer.

**Question 23-10:**   Some have suggested that commercial licenses to catch set amounts of fish should be sold by the federal government, and only those who purchase licenses would be allowed to fish commercially, but the licenses could be traded like stocks on the open market. Critically evaluate that proposal.

**Question 23-11:**   Capture fisheries have been described as "a race to catch the last fish." Discuss whether you agree or disagree with this assessment.

## Aquaculture

Although capture fisheries are in most cases severely overexploited, global fish production is increasing as you saw, as a result of growth of *aquaculture,* farming fish and invertebrates, mainly in China. Two-thirds of aquaculture takes place in inland rivers, lakes, ponds, and artificial tanks. Coastal marine aquaculture (*mariculture*), which includes estuaries and the coastal ocean, accounts for the remainder. The Chinese mainly raise herbivorous species like carp. By contrast, other countries (Norway, Chile, Britain, Canada, and the United States, for example) raise carnivorous species like salmon. Carnivorous species put additional stress on wild fish stocks, since up to 5 kilograms of wild-caught fish (in the form of pellets) must be fed to salmon to produce a kilogram of meat.

Three other concerns arise out of large-scale fish farming: waste production, toxic chemicals, and the deterioration of wild gene stock. Salmon, kept in pens and fed large quantities of fishmeal, produce large quantities of waste.[17] In Norwegian salmon farms, the fish produce as much waste as Norway's 4 million humans. Tests of farm-raised salmon in Scotland and elsewhere have found high levels of PCBs and dioxin, apparently ingested from fish meal, in turn produced from fish taken from the Baltic and other polluted waters. Additionally, farmed salmon are bred for rapid growth and not to develop traits that will aid their survival in the wild. Salmon that escape their pens breed with wild salmon. This could in time reduce their wild cousins' ability to survive in the open ocean. Finally, many farm-raised salmon are fed large doses of antibiotics.

Shrimp mariculture is expanding rapidly as well.[18] Areas developed for shrimp farms are mainly in the coastal tropics, regions under severe strain from the growth of human populations. Shrimp farming is stressful in at least two ways. First, areas devoted to shrimp farming tend to be cleared mangrove swamps (Table 23-2). Mangroves are of critical importance as nurseries for juvenile fish and other marine animals, and the mangrove roots trap and bind sediment washing offshore from rivers. This sediment, if left unchecked, can smother offshore reefs. Reefs, in turn, are important fishing grounds for many subsistence-fishing communities, especially in Southeast Asia.

Coastal areas in Sri Lanka and India where mangroves were cleared were devastated by the 2004 tsunami. Areas with intact mangroves suffered much less damage. Finally, shrimp are fed fishmeal (prepared from wild-caught fish), which, like salmon, puts additional stresses on wild stocks.

If carried out in a manner sensitive to land-use conservation and the preservation of marine fish stocks, aquaculture could be an important addition to a human food base that is becoming threatened by land degradation, pollution, and the addition of 70 million new mouths to feed each year.

The Earth Policy Institute's President Lester Brown is an agricultural expert. Here is what he said recently about the advantages of fish farming:

> Cattle require some 7 kilograms of grain to add 1 kilogram of live weight, whereas fish can add a kilogram of live weight with less than 2 kilograms of grain. Water scarcity is also a matter of concern since it takes 1,000 tons of water to produce 1 ton of grain. But the fish farming advantage in the efficiency of grain conversion translates into a comparable advantage in water efficiency as well, even when the relatively small amount of water for fish ponds is included.[19]

TABLE 23-2 ■ Ecuador: Evolution of Mangroves, Shrimp Ponds, and Saltlands, 1969–1996 (In Hectares)

|  | **1969** | **1991** | **1995** |
|---|---|---|---|
| Area Mangroves | 203,624.00 | 162,186.55 | 149,570.05 |
| Shrimp ponds | 0.00 | 145,998.33 | 178,071.84 |
| Saltlands | 51,752.00 | 6,320.87 | 5,109.47 |
| TOTAL | 255,376.00 | 314,505.75 | 332,751.36 |
| *Source:* www.earthsummitwatch.org/shrimp/national_reports/ecuador. | | | |

---

[17] www.nmfs.noaa.gov; http://www.fao.org/docrep/013/i1820e/i1820e00.htm.

[18] www.h-economica.uab.es/wps/2001_04.pdf.

[19] www.earthpolicy.org.

**Question 23-12:**   List the environmental costs and benefits of aquaculture. Discuss whether you think aquaculture can contribute a sustainable source of food for the world.

**Question 23-13:**   Summarize the major points of this Issue.

## FOR FURTHER THOUGHT

**Question 23-14:**   Bluefin tuna used to be plentiful in the Adriatic Sea. No more. Croatian fisher Lubomir Petricivic said, "You have to work a lot harder to catch fish of any kind. Tuna? Impossible. We don't have any; we can't get it."[20] Simon Cripps of the World Wildlife Fund states, "This is past the alarm stage. We are seeing a complete collapse of the tuna population. It could disappear and never come back."[21] Bluefin tunas are, like most sharks, top predators in their ecosystem. Suggest changes to the food chain if the top predator disappears.

**Question 23-15:**   Go to http://www.scampi.nu/pdf/rap-inter-shrimp-ecuador.pdf, and read a report from the Swedish Society for Nature Conservation on the history of shrimp mariculture in Ecuador. Should some of the farms be "eco-certified?" Why or why not?

**Question 23-16:**   Make a list of the seafood you have eaten in the last year recently, then go to the website www.mbayaq.org/cr/seafoodwatch.asp, click on "seafood search," and fill in the following table with information from that website.

| Seafood Item | Where's It From? | Level of Fishing | More Information |
|---|---|---|---|
|  |  |  |  |
|  |  |  |  |
|  |  |  |  |
|  |  |  |  |
|  |  |  |  |
|  |  |  |  |
|  |  |  |  |
|  |  |  |  |

---

[20] Rosenthal, Aggressive and illegal fishing depletes Mediterranean's tuna, conservationists warn. *New York Times,* 7-16-06, p. A12.
[21] Ibid.

Based on your seafood chart above, analyze whether your diet was seafood friendly and was a part of sustainable fisheries or whether it contained seafood whose capture or culture had an adverse environmental impact.

**Question 23-17:**   Sylvia Earle, a *National Geographic* Explorer-in-Residence and former director of the U.S. National Oceanic and Atmospheric Administration, has said, "People who want to make a difference can choose not to eat fish that are more important swimming alive in the ocean than swimming in lemon slices and butter." Do you agree? Why or why not?

**Question 24-18:**   Research the state of marine capture fisheries. An authoritative source is http://www.fao.org/docrep/015/i2389e/i2389e.pdf. Does the issue as it is presented in this issue paint an objective view of the state of marine fisheries? Discuss.

# Issue 24

## SUSTAINABLE COMMUNITIES:
### *SPRAWL VERSUS SMART GROWTH*

## KEY QUESTIONS

- How do scientists calculate population growth for a locality?
- What is sprawl, and how can we assess its environmental impacts?
- What are the implications of sprawl?
- How do municipalities deal with storm water?
- What is smart growth and what is it designed to accomplish?
- What are brownfields?

## POPULATION GROWTH

"Sprawl is not evil. In fact, it is good. It is the inevitable result of a free people exercising their cherished, constitutionally protected rights as individuals to pursue their dreams when choosing where to live, where to work, where to educate, and where to recreate."[1]

Population growth at the local level involves not only more people, but also high rates of land conversion, often called "development" The type of residential land conversion most common in the United States is *sprawl.* In this type of development, the amount of land converted from its traditional state (e.g., open space, farmland, forests) to residential and commercial use rises at a much faster rate than the growth of the population. For example, a study of urban growth in the greater Charleston, South Carolina metropolitan area from 1973 to 1994 found that over the twenty-one-year period, the rate of urban land use growth exceeded population growth by a 6:1 ratio.[2]

L. Brooks Patterson is certainly entitled to his opinion, but sprawl is associated with numerous kinds of environmental issues, not the least of which is transport emissions, especially $CO_2$. In a report done by the World Bank cities were ranked according to their population densities and $CO_2$ transport emissions per capita per year. Here is a brief snapshot of the findings.[3]

---

[1] The county executive of Oakland Co., MI, L. Brooks Patterson, http://www.oakgov.com/exec/brooks/sprawl.html.

[2] Allen, J., & Kang Shou Lu. Modeling and predicting future urban growth in the Charleston area. The Strom Thurmond Institute for Government and Public Affairs, Clemson University. www.strom.clemson.edu/teams/dctech/urban.html.

[3] *The Economist*, Shoots, greens and leaves; June 16, 2012, p. 69.

| City | Population Density/ Hectare | CO₂ Transport Emis- sions Per Capita/Y* |
|------|----------------------------|----------------------------------------|
| Atlanta | 2.4 | 7.5 |
| Houston | 14 | 6.2 |
| Melbourne | 16 | 2.5 |
| London | 50 | 1.2 |
| Barcelona | 160 | 1.6 |
| Bangkok | 185 | 0.8 |

Note: Barcelona and Atlanta have similar populations.
*Tonnes

**Question 24-1:** Describe the association between urban population density and transport emissions.

**Question 24-2:** Atlanta's metropolitan area is home to about 5.2 million people. How many tonnes of $CO_2$ per year could be saved if Atlanta was as efficient as Barcelona, with a similar population?

Prince William County, Virginia, located 35 kilometers south of Washington, D.C. along Highway I-95, has been one of the more growth-oriented entities in its region. From 1940 to 1990, Prince William's population grew at the rate of 4 percent per year. From 1990 to 2000, growth rates decreased to 2.6 percent. Its population in 2010 was 402,000 with a population density of 1,195 persons per square mile.[4] We will use Prince William County as our first case study of local growth. Between 2000 and 2007 alone, 9.4% of the entire county area was developed from open space/forest to commercial/residential structures.

## GROWTH IN PRINCE WILLIAM COUNTY

Recall from Issue 1 (page 31) that when a population grows exponentially, the time it takes for the population to double, called doubling time, can be calculated using the formula $t = 70/r$, where t equals doubling time (usually in years) and r is the growth rate expressed as the percentage increase or decrease (for example, a rate of 7%, or 0.07, would be entered as 7). Doubling time is a very useful and practical concept for *projecting* (not *predicting*) and analyzing the implications of growth.

---

[4] http://quickfacts.census.gov/qfd/states/51/51153.html.

**Question 24-3:** Population growth from 2000 to 2010 was 43.2%.[5] What then was the average rate of growth over that ten-year period?

**Question 24-4:** The 2010 population of Prince William County was 402,000. Using the doubling time formula, determine approximately when the population of Prince William County would reach 1 million. Starting with the 2010 population, use the most recent percent annual rate (2000–2010).

**Question 24-5:** Now, let's apply the doubling time formula to project land-use changes. The area of Prince William County is approximately $9 \times 10^8$ m$^2$. At a 4.3 percent growth rate, when would Prince William County reach a population density of 1 person/100m$^2$?

**Question 24-6:** Answer the question again, using an earlier growth rate of 2.6 percent. Compare the two results and state how much more time it would take Prince William County to reach this high density with the lower growth rate.

**Question 24-7:** Discuss to what extent lowering the growth rate significantly reduces the impact of population growth.

---

5 www.pwcgov.org.

# IMPERVIOUS SURFACE

Development is an inclusive and somewhat ambiguous term used to refer to all the human-built structures in an area. One major effect of development is the increased amount of *impervious surface*. Impervious surface refers to any surface material that water cannot penetrate. In this section, we will analyze impervious surface as a local population growth issue.

When most people speak of development, they mean subdivisions, commercial buildings (such as offices, shops, and malls), and roads. Most of these buildings have parking areas attached. The parking area can be as small as the driveway of a house or it can be a paved area adjacent to a large mall, covering tens of thousands of square meters. These impervious surfaces collect runoff and prevent it from infiltrating into soils and surface sediment, where rainfall can be stored and natural filtering can often remove some pollutants.

Paved areas, and the vehicles that are parked on them, can contribute significant amounts of pollution to the water. But even if the runoff were to contain no pollution, it still would increase the risk of flash flooding. Local government officials are familiar with these threats and try to design stormwater management systems to handle runoff from development. Most municipal systems consist of a network of pipes that collect runoff from streets and large parking lots and channel it into artificial stormwater detention ponds or into creeks in the vicinity. Sometimes, the runoff pipes carry the water to a sewage treatment plant, where it can be treated before it is discharged into streams.

Stormwater management systems can, but rarely do, incorporate *Better Site Design* principles, including reshaped zoning regulations, increased green space, pervious concrete, vegetated swales and buffers, rain gardens, narrower streets, and even vegetated rooftops.

Although treating runoff to remove pollutants is a good idea, it can cause serious problems during times of heavy runoff. The added runoff from the stormwater system can, and often does, overload the sewage treatment plants, resulting in a mixture of untreated sewage and storm water dumped into waterways. For this reason, many communities design separate systems to transmit sewage and stormwater. Although this is more expensive, it results in much less environmental degradation, especially during floods. Unfortunately, it can also result in polluted water from impervious surfaces contaminating waterways.

Stormwater management systems work only as well as the weakest link in the chain, which simply means that if anything can go wrong, it usually will. Pipes and retention ponds may get clogged with debris or fill up with vegetation and sediment. Storm drains may be too small to handle runoff volumes, and development since construction of the system may overload them. This means that unless local governments take painstaking care in the design and maintenance of stormwater management systems, a costly endeavor that cash-strapped cities often forego, the more development in an area, the higher the risk of flash flooding.

# HURRICANES AND TROPICAL STORMS

A hurricane or tropical storm can often dump as much rain on an area in a few hours as the area experiences in several months. Consider this example from Prince William County. The Potomac Mills Mall is one of the largest in the United States, and in 1999 the buildings and parking area together represented 5,675,411 ft$^2$ of impervious surface.[6]

When the last great hurricane, Agnes, hit the area in June of 1972 and dumped 14 inches of rain on the region in three days, Agnes' effects were described by officials as

---

[6] Prince William County Department of Public Works (Prince William County did not publish impervious surface data for 2000 and later.)

the Eastern Seaboard's most costly disaster in history. But Potomac Mills Mall hadn't yet been built.

**Question 24-8:** Assume another Hurricane Agnes-size storm occurs (as it surely will someday) and dumps 14 inches of rain on the region in 72 hours. Assume further that before Potomac Mills Mall was built the land was pasture and forest, which, given the area's subsurface geology, could have absorbed much of the storm's runoff. Calculate the maximum amount of extra runoff Potomac Mills Mall would generate when the next "Hurricane Agnes" hits northern Virginia. Express your answer in cubic feet and liters ($1 \text{ ft}^3 = 2.83 \times 10^{-2} \text{ m}^3$; $1 \text{ m}^3 = 1,000 \text{ L}$).

**Question 24-9:** Assume this runoff has to be handled over a 72-hour period. What is the resultant discharge from Potomac Mills Mall in liters per hour? (For comparison, a typical summer flow of the Potomac River is 100 million L/hr.)

Of course, this number represents a maximum value, but it illustrates how development can increase a region's susceptibility to flash flooding. Keep this point in mind while you study the impact of home building on Prince William County's impervious surface in the following section.

## HOME BUILDING AND IMPERVIOUS SURFACE

Table 24-1 contains data on housing units and impervious surface for Prince William County from 1940–2010.[7] The area of the county is 222,615 acres or 90,000 hectares.

TABLE 24-1 ■ Prince William County's Housing Data Impervious Surface Area[1]

| Year | Housing Units | Impervious Surface Area ($m^2$) | Area of county covered by impervious surface (from housing alone) % |
|------|---------------|--------------------------------|--------------------------------------------------------------------|
| 1940 | 3,545 | 622,018 | 0.07 |
| 1950 | 5,755 | 1,009,850 | 0.11 |
| 1960 | 13,207 | 2,317,519 | 0.25 |
| 1970 | 29,885 | 5,244,050 | 0.58 |
| 1980 | 46,490 | 8,157,855 | 0.90 |
| 1990 | 74,759 | 13,107,552 | 1.45 |
| 1994 | 90,759 | 15,175,716 | 1.68 |
| 2000 | 94,570 | — | — |
| 2002 | 108,004 | — | — |
| 2005 | 123,379 | — | — |
| 2010 | 137,115 | — | — |
| [1]Source: Prince William County Department of Public Works. | | | |

[7] Prince William County Department of Public Works.

**Question 24-10:**   On the first set of axes below, plot housing units versus time. On the second set, plot impervious surface area (m²) versus time. (Use data from 1940–2010.) Interpret the graphs.

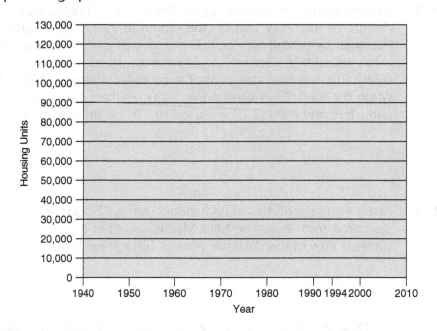

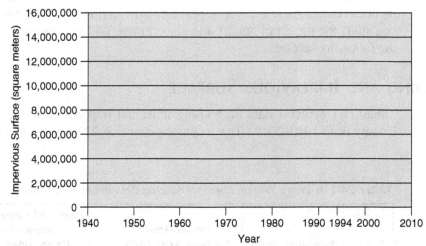

(To answer the following questions, use the equation, $r = (1/t)\ln(N/N_0)$, explained in "Using Math in Environmental Issues," pages 6–8.)

**Question 24-11:**   Calculate average annual growth rates for both housing units and impervious surface area for the period 1940–94. How do they compare?

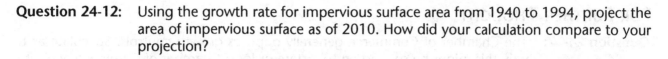

**Question 24-12:** Using the growth rate for impervious surface area from 1940 to 1994, project the area of impervious surface as of 2010. How did your calculation compare to your projection?

**Question 24-13:** Speculate how life would be different in Prince William County if half the county were impervious surface.

**Question 24-14:** In Question 24-4, we asked you to project when Prince William County's population would reach a density of one person per 100 square meters, which is certainly a very high density. It has been observed that population growth will stop eventually. We can decide when, or we can let math and "nature" take its course. Comment on why no one seems to be planning for extremely high population densities.

**Question 24-15:** Discuss whether "sustainable growth" is an oxymoron.

**Question 24-16:** Summarize the main points of this Issue.

# FOR FURTHER THOUGHT

**Question 24-17:**   The Chamber of Commerce generally opposes growth controls. Speculate as to why this might be so. Log on to a state or local Chamber of Commerce website, and then summarize and critically evaluate their position on growth controls.

**Question 24-18:**   Log onto the Sierra Club's website (www.sierraclub.org/sprawl/). How do they explain their position on sprawl and local population growth? Critically assess it.

**Question 24-19:**   Advocates of unrestricted growth frequently argue that growth expands the tax base, which lowers the tax burden for everyone and thus improves the quality of life. However, the results of more than seventy community studies conducted by the American Farmland Trust indicate for every dollar of revenue generated by residential development, local governments incur a median cost of $1.15 to provide services, and costs can be much higher. By comparison, the median cost of providing municipal services to farm, forest, and open land is $0.35 per dollar of revenue generated. The median cost of providing services to commercial/industrial development is $0.27 per dollar generated.[8] Research and discuss this issue.

**Question 24-20:**   Find U.S. Census data (www.census.gov) or go to another source for data on growth for your county and make similar calculations to those we did for Prince William. (The California Department of Finance has excellent statistics, for example.)[9] What is the growth rate? What are its implications?

**Question 24-21:**   Consider transportation. In 2012, there were 756 motor vehicles per 1,000 persons in the United States. Assuming the number of motor vehicles grows with the population, assess the impact of population growth on road congestion in your county or city. Is road construction and maintenance an unavoidable cost of population growth? Should people who choose not to own cars pay for these new or expanded roads? Provide evidence for your conclusions.

**Question 24-22:**   Examine your local zoning ordinances, which you can usually find at your city or town planner's office. Are they designed to promote or limit growth?

**Question 24-23:**   Study articles from your local newspaper or from a metropolitan newspaper in your library that focus on growth. Are growth rates given? If not, can you calculate them? Do the articles describe growth as "healthy" or "robust"? Do you agree that growth is "healthy" or "robust"? Discuss. Many of these articles refer to high population growth as "population booms" or "baby booms." What is the connotation of these descriptions? Can you think of a term that is neutral or more appropriate?

**Question 24-24:**   To find out what your state is doing about growth, go to www.smartgrowth.org/. Discuss whether your state is taking actions to promote or discourage growth.

---

[8] www.farmland.org.
[9] California Department of Finance, www.dof.ca.gov.

# SUSTAINABLE COASTAL DEVELOPMENT

## KEY QUESTIONS

■ What are the options for "developing" the U.S. coastline?
■ What are the costs and benefits of these options?
■ Who are the stakeholders in coastal development?
■ Who decides how the coastline will be developed?

## COASTAL POPULATION GROWTH

Beaufort (pronounced *BYEW-fert*), South Carolina, is known as *beautiful Beaufort* because of its extensive networks of salt marsh-lined creeks and moss-draped oak trees, hallmarks of the South Carolina lowcountry. The natural beauty of the area combined with the relaxed pace of life has attracted transplants, predominantly retirees from the Northeast, and this has resulted in a building boom. From 1990 to 2000, the population of Beaufort County grew from 86,425 to 120,937, an increase of 39.9 percent. By 2010, it had increased to 162,900.

**Question 25-1:** By what annual percent did the population increase between 1990 and 2010?

**Question 25-2:** Based on this rate, when would Beaufort County's population double (recall the doubling time formula used throughout this book)?

In North Carolina, coastal Brunswick County grew even faster over the same period, from 50,985 to 73,143 (43.5 percent). By 2010 it had reached 107,431. As fast as Beaufort and Brunswick counties are growing, they can't compare to the nation's fastest growing county, Flagler County, on Florida's northern coast, which grew 10.7 percent from 2004 to 2005 alone.[1] By 2010, it had reached 95,696.

---

[1] All population numbers, unless otherwise stated, are from www.census.gov.

Coastal areas of the United States make up about 17 percent of the country's land area, but they hold about 53 percent of the total population. Of the twenty-five most densely populated counties in the United States, twenty-three are coastal. All regions of the coastal United States (Northeast, Southeast, Gulf of Mexico, Pacific, and Great Lakes) are expected to grow faster than the national average. The Southeast, which in 2003 was the least populated coastal region in the United States, is expected to grow the fastest, hurricanes to the contrary. Some Florida and North Carolina counties are projected to grow by as much as 16 to 17 percent over that period. Population densities are highest in the Northeast coastal region. Densities increased over the period 2003 to 2008 from 641 to more than 660 persons per square mile.

**Question 26-3:** There are 260 hectares in a square mile. What is the northeast coastal density expressed in persons per hectare? Compare this density to that of Houston, TX, at 14, from the last Issue, or London, at 50.

**Question 25-4:** Growth of the magnitudes presented above has been described as a mixed blessing. What do you think this means? Be specific and use examples.

**Question 25-5:** Carl Laundrie, a Flagler County, Florida, administration spokesman, said, "There isn't a county in the world that's prepared for 10 percent growth" a year.[2] Discuss ways in which a county might be unprepared for high growth rates.

In Issue 3, you analyzed and evaluated population growth in Bangladesh, an impoverished country vulnerable to natural disasters. In this Issue, you will do the same for the Southeast coastal region. Using a case study developed by the National Oceanic and Atmospheric Administration's (NOAA) Coastal Services Center, you will study three options to develop a small coastal area in Georgia. Then, you will analyze the environmental, economic, and social impacts of these developments.

## THE COASTAL GROWTH MODEL

The NOAA Coastal Services Center's model is called *Alternatives for Coastal Development: One Site, Three Scenarios.*[3] The site used in the model, a peninsula surrounded by

---

[2] www.wilmingtonstar.com/apps/pbcs.dll/article?AID=/20060316/NEWS/60315042/-1/frontpage.
[3] The original website, www.csc.noaa.gov/alternatives, is no longer functional. However, an archived version is at http://web.archive.org/web/20090307215818/http://csc.noaa.gov/alternatives/.

tidal creeks and marshes, is a real 1,100-acre location in Georgia that is currently being privately developed. The three development designs are Conventional (*Point Peter Estates*), Conservation (*Point Peter Preserve*), and New Urbanist (*Point Peter Villages*) (Figure 25-1).

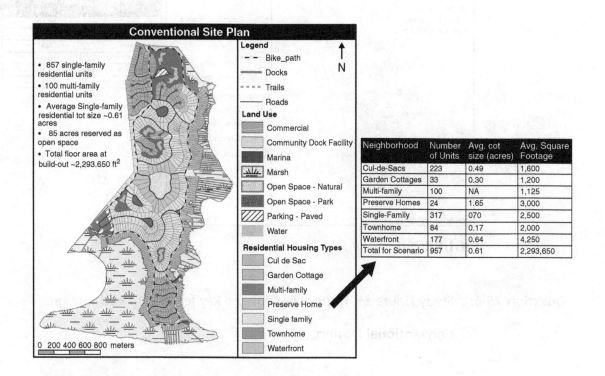

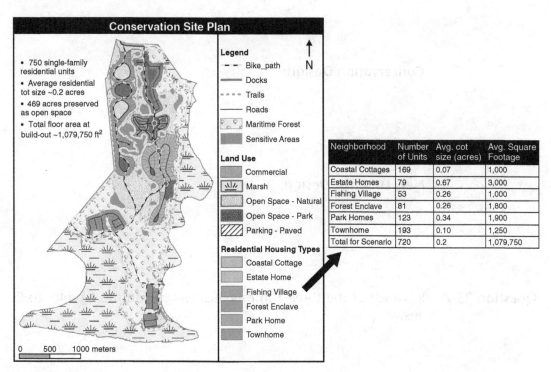

**FIGURE 25-1** The three design models for the development of Point Peter Peninsula. (www.csc.noaa.gov)

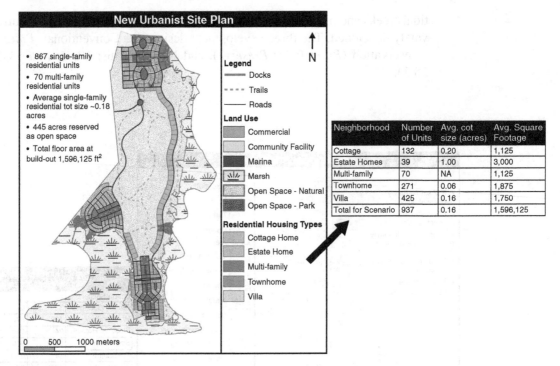

**FIGURE 25-1** *(cont.)*

**Question 25-6:** Study Figure 25-1, then describe the key features of each design.

**Conventional Design:**

**Conservation Design:**

**New Urbanist Design:**

**Question 25-7:** In which of the three communities would you most prefer to live? Explain your answer.

TABLE 25-1 ■ Environmental indicators of the three coastal development alternatives

| Indicator | Conventional | Conservation | New Urbanist | Description |
|---|---|---|---|---|
| **Open Space** | | | | Amount of land left undeveloped (natural) or as parks, playgrounds, etc. (managed). |
| Percent of Site | 15 % | 71 % | 67 % | |
| Total Acres | 85 | 469 | 445 | |
| Natural acres | 28 | 432 | 403 | |
| Managed acres | 57 | 37 | 42 | |
| **Docks** | | | | Total linear feet of all docks and total area covered by docks. |
| Total dock length (feet) | 43,721 | 1,013 | 2,086 | |
| Total dock area (feet$^2$) | 262,326 | 6,078 | 12,518 | |
| **Paths, Trails, and Sidewalks** | | | | Total length of trails with the total broken into total pervious and impervious lengths |
| Total (linear feet) | 32,159 | 78,846 | 127,134 | |
| Impervious | 32,159 | 26,946 | 78,859 | |
| Pervious | 0 | 51,900 | 48,275 | |
| **Water Consumption** | | | | Estimated water consumption totals for residential and park parcels based on national and local averages |
| Estimated total gallons/day | 358,926 | 231,584 | 287,765 | |
| **Impervious Surface** | | | | Amount of impervious surface. |
| Percentage (of total site) | 26 % | 12 % | 18 % | |
| Total Acres | 169 | 82 | 119 | |
| **Pollutant Runoff** | | | | Results are the change between the amount of runoff from the undeveloped site and the amount of runoff from the site if developed. |
| Change in total runoff volume to surrounding waters | 53.13 % | 28.28 % | 42.39 % | |
| Change in total nitrogen | 153.32 % | 84.46 % | 114.39 % | |
| Change in total phosphorus | 640.07 % | 297.79 % | 357.35 % | |
| Change in total suspended solids | 61.16 % | 43.31 % | 68.86 % | |

*Source:* Adapted from www.csc.noaa.gov/alternatives/environmental.html.

The three communities have different environmental impacts. Table 25-1 provides an overview of some of the environmental features of each design.

**Question 25-8:** Which of the three communities has the lowest environmental impact? Cite evidence from Table 25-1 to support your answer.

Economically, the New Urbanist design was projected to be the most profitable, due to a number of factors including lower construction and infrastructure costs and increased property values arising out of the exclusivity of the development.

**Question 25-9:** One widely accepted definition of sustainability encompasses the triple bottom line of people, planet, and prosperity (sometimes listed as *profit*). Which of the three designs is the most sustainable, according to this definition? Which is the least? Explain the reasoning behind your answer.

**Question 25-10:** Discuss whether any of these developments can be called sustainable considering that each replaces natural landscape with a built environment and that many residences in this type of development are second, vacation homes.

**Question 25-11:** This model you just examined is hypothetical, yet it is based on a real area that is being developed. Land Resources Companies is calling its development Cumberland Harbour, and promotional material on its website implies that the company is using a conventional design.[4] Based on your analysis of the three development alternatives, in the space below compose a letter to the developer either supporting or opposing their design. Support your decision with evidence.

---

[4] http://www.cumberlandharbourlots.com/. Click on the *satellite photo* or *site map* buttons on the sidebar to see an aerial or map view of the development.

**Question 25-12:** Summarize the major points of this Issue.

## FOR FURTHER THOUGHT

According to Environmental Defense, one way to characterize whether a community is in "balance" is to compare jobs to residents:

> A community with a healthy balance of jobs and housing is one where people can both work and afford to live. With this balance comes the possibility of fewer people commuting, and less regional traffic congestion and air pollution. Municipalities that fail to achieve a balance of jobs and housing do so for many reasons. It can happen when a local government allows residential development without adequate commercial development, or when residential development is limited but commercial development is encouraged.[5]

Accounting for children and those not looking for work, a ratio of 0.6 to 0.85 (i.e., 0.6 to 0.85 jobs per resident) is believed to indicate a "balanced" community. Lacking an integrated mass-transit system (which is missing in the Denver area—see next question), higher values will bring unwanted auto commuters into the community, and lower values will drive residents elsewhere to look for work. So, imbalance will unnecessarily add road costs, which residents are increasingly unwilling to pay for.

**Question 25-13:** Go to the American Planning Association report "Jobs-Housing Balance" at http://www.planning.org/pas/reports/subscribers/pdf/PAS516.pdf, and scroll to page 19. Why is Longmont, in the Denver area, considered a model for other cities in the region? Is Boulder considered a balanced community? Why or why not? What is Boulder doing to address the issue?

**Question 25-14:** Go to the website of the Noisette Company (see Issue 26; http://navyyardsc.wordpress.com/). Discuss the Noisette Company's development philosophy and approach from the perspective of sustainability and sustainable communities.

**Question 25-15:** Research the topic of low-impact development. Summarize your findings in a paragraph or two.

**Question 25-16:** Research how New Orleans is being rebuilt from Hurricane Katrina, by referring to the website of Rebuilding Together New Orleans, http://www.rtno.org, or the local newspaper, the *Times Picayune*, http://www.timespicayune.com/. Summarize your findings and comment on the sustainability of the plans.

---

[5] www.environmentaldefense.org/documents/1302_DenversprawlReport.pdf.

# Issue 26

## SUSTAINABLE BUILDINGS AND HOUSING

### KEY QUESTIONS

- What is "high performance green building"?
- Does a "green" building cost more than a conventional one?
- What is an EarthCraft House?
- What does LEED-certified mean?
- Can buildings be made truly sustainable?

### INTRODUCTION: THE NOISETTE PROJECT[1]

On and around the site of the old Navy Base in North Charleston, South Carolina, one of the biggest green urban-redevelopment projects in the United States is underway. Noisette, named after the noted eighteenth century botanist Philip Noisette, is a 3,000-acre "sustainable community" being developed as a public-private partnership between the City of North Charleston and the Noisette Company. When completed in about twenty years, the $1 billion project will include over 4,000 housing units, a waterfront park, a 200-plus acre tidal creek preserve, and about 5 million square feet of retail, industrial, and commercial space. The project's philosophy of sustainability is based on the *Sanborn Principles* (Figure 26-1) developed by urban designers in the 1990s, which is an approach that considers social, economic, and environmental well-being.

When planning for the project commenced, the median income of North Charleston residents was 50 percent of the mean income of South Carolina, one of the nation's poorer states. Property values had been flat or declining for at least two decades. The area's business district was 70 percent vacant. The high school drop out rate was 70 percent.

In the 1950s and 1960s, North Charleston was a bustling industrial and military hub, with unregulated smokestacks and effluent pipes churning out polluted air and viscid, toxic fluids 24 hours a day. A veritable laboratory of environmental health and justice issues, most of North Charleston was not a desirable location to live if you could afford to live elsewhere. Hazardous wastes, including solvents, paint strippers, lubricants, and heavy metals, were either stored or disposed of improperly and often found their way into the soil and into waterways. The environment, including what had been a garden city park designed by the Olmsted brothers (designers of Central Park) in the late 1800s, was paved over or otherwise heavily degraded.

The built environment of the area Noisette was developing was heavily run-down. Commercial buildings were unoccupied and crumbling, and residential housing, mostly

---

[1] http://navyyardsc.wordpress.com/

**Ecologically Responsive:** The design of human habitat shall recognize that all resources are limited and will respond to the patterns of natural ecology. Land plans and building designs will include only those with the least disruptive impact upon the natural ecology of the Earth. Density must be most intense near neighborhood centers where facilities are most accessible.

**Healthy, Sensible Buildings:** The design of human habitat must create a living environment that will be healthy for all its occupants. Buildings should be of appropriate human scale in a non sterile, aesthetically pleasing environment. Building design must respond to toxicity of materials, lighting efficiency and quality, comfort requirements, and resource efficiency. Buildings should be organic and integrate art, natural materials, sunlight, green plants, energy efficiency, low noise levels, and water. They should not cost more than current conventional buildings.

**Socially Just:** Habitats shall be equally accessible across economic classes.

**Culturally Creative:** Habitats will allow ethnic groups to maintain individual cultural identities and neighborhoods while integrating into the larger community. All population groups shall have access to art, theater, and music.

**Beautiful:** Beauty in a habitat environment is necessary for the soul development of human beings. It is yeast for the ferment of individual creativity. Intimacy with the beauty and numinous mystery of nature must be available to enliven our sense of the sacred.

**Physically and Economically Accessible:** All sites within the habitat shall be accessible and rich in resources to those living within walkable (or wheelchair-able) distance.

**Evolutionary:** Habitats' design shall include continuous re-evaluation of premises and values, shall be demographically responsive, and flexible to change over time to support future user needs. Initial designs should reflect our society's heterogeneity and have a feedback system.

**FIGURE 26-1** The Sanborn Principles for urban sustainability. (The Sanborn Principles were developed at a conference organized by Barbara Harwood with funding by the National Renewable Energy Laboratory in 1994. Participants in the conference included Harwood, Perry Bigelow, Bill Browning, Oliver Drerup, Liz Gardener, John Knott, Amory Lovins, Paul MacCready, Ned Nisson, Richard Register, and others. They are currently being used as guidance for urban development in several countries around the world.)

post–World War II vintage designed as temporary residences, was deteriorating. A centerpiece of the Noisette project is their plan to refurbish older structures and build new buildings and homes sustainably. Accordingly, Noisette adapted green building techniques employed elsewhere to fit the Southeast coastal climate.

What is sustainable building? What are green building techniques? We will examine these in this issue.

## HIGH-PERFORMANCE BUILDING

High-performance building, also known as sustainable or green building, refers to making structures that are durable, are energy and water efficient, and have high indoor air quality. They are made of materials with low environmental impact. Construction protects the natural surroundings and minimizes waste.

Commercial and residential buildings have a considerable environmental impact. According to the U.S. Green Building Council and the U.S. Department of Energy, constructing and operating buildings in the United States accounts for

- nearly two thirds of total U.S. electricity consumption,
- a third of U.S. greenhouse gas emissions,
- nearly half of sulfur dioxide emissions,

- a quarter of nitrous oxide emissions,
- 10 percent of particulate emissions,
- 35 percent of the country's carbon dioxide emissions,
- 136 million tonnes of construction and demolition waste (see Issue 15),
- 12 percent of potable water use (see Issue 11), and
- 40 percent (3 billion tons annually) of global raw materials use.

Conventionally built structures typically use building materials that are a part of the *take-make-waste* linear extraction process. These materials begin as natural resources (trees, minerals) that are extracted, processed, refined, manufactured, and transported using mainly fossil energy, generating waste and pollution at each step before ultimately winding up in a landfill.

In addition, conventional buildings are designed to last only thirty years. Furthermore, they negatively impact the health, productivity, and learning of occupants. One study estimated that poor indoor environmental quality in commercial buildings results in as much as $168 billion in productivity losses annually.

**Question 26-1:** How would you describe the air quality in your current location? Do you smell any chemical odors, or does the air smell fresh and clean? Does being in the structure give rise to any allergies on your part or any of your colleagues?

## LEED AND EARTHCRAFT CERTIFICATION

The U.S. Green Building Council's LEED certification program is the standard rating system for green commercial buildings. LEED, which stands for *Leadership in Energy and Environmental Design,* is a point-based system that certifies buildings according to the degree of sustainability, progressing from the most basic LEED-certified, to silver, gold, and the pinnacle of sustainability, platinum. By 2011, more than 44,000 LEED-certified buildings in government, business, industry, and education were either completed or under construction in all fifty states and in thirteen foreign countries.

Currently, there are three major regional or national certifications for high performance green homes, along with numerous local programs. EarthCraft House, designed specifically for the climate of the Southeast, is a partnership between the Greater Atlanta Home Builders Association, Southface Energy Institute, and government and industry partners. LEED for Homes is a program of the U.S. Green Building Council for homes. The National Association of Home Builders (NAHB) offers *Green Home Building Guidelines*. In addition to these, the U.S. Department of Energy's Energy Star program certifies homes that meet energy-efficiency standards.

## ECONOMICS OF GREEN BUILDING

A persistent claim about green building is that it costs substantially more than conventional building. Let's examine that claim. First, we'll consider what builders call *first costs,* the actual cost to develop the site and construct the building (but not the cost of the land). Several comprehensive LEED cost studies have been conducted, including California's Sustainable Building Study, the Davis Langdon Study, and the GSA LEED Cost Study.

The results: up-front costs for green buildings typically ranged from 0 to 2 percent more than for conventional buildings. There was some variability. LEED platinum buildings had the highest premium, whereas LEED libraries cost less to build than conventional counterparts.

**Question 26-2:** What is the cost per square foot if the premium for constructing a LEED Silver building is 2 percent greater than the $150 per square foot cost of a conventional building?

**Question 26-3:** How much more would a 25,000-square-foot LEED-certified building cost compared to a conventional building, assuming a 2 percent cost increase for the LEED building?

**Question 26-4:** In 2005, the latest year for which we have data, there were approximately $5 \times 10^6$ commercial buildings comprising $75 \times 10^9$ square feet of floor space in the United States. First, calculate the combined construction costs for these buildings, assuming $150 per square foot, and then recalculate as if the buildings had been LEED-certified (i.e., the cost was 2 percent higher).

**Question 26-5:** Repeat the calculations for the $5 \times 10^6$ ft$^2$ of commercial, retail, and warehouse space in the Noisette Project (assuming it is all new construction).

First costs are important, but they are only one measure of a building's true cost. Figure 26-2 shows other costs for buildings in the United States. The California Energy Commission, the Rocky Mountain Institute, and others (including the U.S. Postal Service) have shown that green buildings provide benefits to occupants that increase their productivity and decrease absenteeism. A study by the Lawrence Berkeley National Laboratory estimated U.S. businesses could save as much as $58 billion in lost sick time and an additional $200 billion in worker performance annually by simply improving indoor air quality.[2]

[2] Fisk, W. J. 2000 "Health and Productivity Gains from Better Indoor Environments and their Relationship with Building Energy Efficiency." *Annu. Rev. Energy Environ.* 25: 537–566.

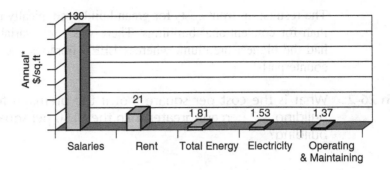

**FIGURE 26-2** Costs, excluding first costs, associated with owning and operating a building in the United States. (*1991 Source: BOMA, EPRI, Statistical Abstract in RMI "Greening the building and the Bottom Line")

**Question 26-6:** Assume that salary and benefits combined amount to $200 per square foot in 2006.[3] What is the annual economic benefit for the following cases:

a. Employee productivity increase = 1 percent for a 25,000-square-foot building

b. Employee productivity increase = 10 percent for a 25,000-square-foot building

c. Employee productivity increase = 1 percent for $75 \times 10^9$ square feet of floor space (total for United States)

d. Employee productivity = 10 percent for $75 \times 10^9$ square feet of floor space in the United States

**Question 26-7:** For a 25,000 square foot building over a 30-year span, calculate which costs more to build and staff, a conventional building or a LEED-certified building (assuming a 1 percent employee productivity increase).

---

[3] The data from Figure 26-2 are from 1991 and do not include benefits. Thus, a better estimate of salary costs in 2006 is closer to $200 or more.

**Question 26-8:** Based on your calculations above, explain whether you would build a conventional or a LEED-certified building given the opportunity.

In addition to improved worker productivity, there is evidence of substantial educational benefits of green building. A California Energy Commission study of performance of over 24,000 elementary school students reported that "The physical conditions of classrooms, most notably the window characteristics, were as significant and of equal or greater magnitude as teacher characteristics, number of computers, or attendance rates in predicting student performance."[4] In classrooms with daylighting, a feature of most LEED-certified buildings, standardized test scores increased from 7 to 18 percent. Other studies have shown improved student behavior and longer attention spans in LEED-certified buildings.

Other benefits of commercial green building include reduced utility (electricity, gas, and water) costs, reduced greenhouse gas emissions, improved performance of mechanical systems (heating, air conditioning, etc.), and more durable and longer-lasting structures. A study by the consulting firm Capital E estimated that the twenty-year net benefit of green building, considering first costs and productivity gains, was $50 to $65 per square foot, depending on the level of LEED certification.[5] In addition, the use of materials with low environmental impact results in less use of virgin natural resources and greater preservation of ecosystems (see Figure 26-3).

**Question 26-9:** Explain whether LEED-certified buildings cost more than conventional buildings.

---

- High recycled content and recyclable
- Locally produced materials (thus avoiding environmental impacts of transportation)
- Low embodied energy (takes lower amounts of energy to produce)
- Durable
- Low environmental impact
- Energy and water saving
- Improved indoor air quality
- Nontoxic
- Renewable
- Standard sizes, modular, precut (reduces waste)
- Bio-based

---

**FIGURE 26-3** Characteristics of sustainable materials. (Reprinted with permission from the Responsible Purchasing Network (RPN), Center for a New American Dream. RPN helps institutional purchaser's identify and select socially and environmentally preferable goods and services. www.responsiblepurchasing.org)

---

[4] http://www.energy.ca.gov/2003publications/CEC-500-2003-082/CEC-500-2003-082-A-07.PDF
[5] www.cap-e.com.

## Residential Green Building

Residential green (high performance, sustainable) building embodies the same general features of commercial green building, except on a smaller scale. We are going to restrict our analysis to a single feature—energy efficiency.

In 1940, the typical new American house was about 950 ft$^2$. In 2011, the average new American house was about 2,480 square feet, even though household size has dropped from 3.7 in 1940 to 2.63 over the same period.

**Question 26-10:** Calculate the area per family member for an average U.S. house in 1940 and 2011.

**Question 26-11:** Discuss to what extent you think the additional space is justified, desirable, or necessary.

Compared to smaller houses, large houses (the largest varieties called "McMansions" by critics) use more material resources and energy and have a large environmental footprint. The U.S. National Association of Home Builders (NAHB) estimates that construction of a 2,082 ft$^2$ house uses 13,837 board-feet of framing lumber, 11,550 ft$^2$ of sheathing, 2270 ft$^2$ of flooring, 3103 ft$^2$ of roofing, and 16.92 tonnes of concrete.

**Question 26-12:** How much lumber, sheathing, flooring, roofing, and concrete would a 2,500 ft$^2$ house use (assuming the same proportions as the 2,082 ft$^2$ house).

In 2005, the U.S. Energy Information Administration released a report on electricity consumption for the year 2001. Heating, ventilation, and cooling (HVAC) accounted for 356 billion kWh, representing 31 percent of the electricity consumed by households. Kitchen and laundry appliances accounted for about 305 billion kWh. (Refrigerators consumed the most among household appliances—156 billion kWh of electricity). Home electronics used an additional 82 billion kWh. For 2001, the household electricity intensity, defined as the annual electricity consumed per household, was 11,600 kWh per household. There were approximately 114 million households in 2010.[6]

According to the U.S. Green Building Council and the Southface Energy Institute, constructing a house with a tightly sealed thermal envelope, a high efficiency HVAC

---

[6] www.census.gov/prod/1/pop/p25-1129.pdf.

system, good insulation, Energy Star appliances, and proper orientation, increases energy efficiency by about 40 percent compared to a conventionally built house. For a $250,000 home in the Southeast, the cost of achieving this level of energy efficiency is about $3,000.

**Question 26-13:** Calculate the household electricity intensity (kWh per household) for a 2001 house built with the above features, which increase efficiency by 40 percent.

**Question 26-14:** Calculate the energy savings in kWh and dollars (ignoring the $3,000 capital cost and assuming the cost of electricity as $0.10/kWh) if all 114 million households occupied energy efficient houses.

**Question 26-15:** Ignoring inflation, in a typical 2,500 ft$^2$ house, what is the payback time for the initial $3,000 investment if the electricity cost is $0.10/kWh? Assume electricity intensity of 11,600 kWh per year.

**Question 26-16:** Discuss to what extent you think it makes sense to install high-efficiency energy improvements in new homes.

**Question 26-17:** Why do you think all new homes are not automatically made as energy efficient as possible?

Depending on the source of your electricity (coal, natural gas, etc.), saving money on your electric bill may also result in reduced emissions of $CO_2$ at the power plant supplying your house. Table 26-1, from the EPA, gives average $CO_2$ emission factors (in pounds of $CO_2$ equivalents released per kWh produced).

**Question 26-18:** Find your state in Table 26-1. Was its $CO_2$ emissions high or low compared to the other states? Explain your observation.

**TABLE 26-1** ■ $CO_2$ emissions factors (in units of pound of $CO_2$/kWh) for U.S. states in 2007 (EPA).

| State | $CO_2$/kWh | State | $CO_2$/kWh |
|---|---|---|---|
| Alabama | 1.30 | Montana | 1.57 |
| Alaska | 1.11 | Nebraska | 1.50 |
| Arizona | 1.22 | Nevada | 1.57 |
| Arkansas | 1.28 | New Hampshire | 0.78 |
| California | 0.70 | New Jersey | 0.71 |
| Colorado | 1.99 | New Mexico | 1.99 |
| Connecticut | 0.75 | New York | 0.91 |
| Delaware | 1.80 | North Carolina | 1.22 |
| DC | 3.61 | North Dakota | 2.39 |
| Florida | 1.35 | Ohio | 1.78 |
| Georgia | 1.39 | Oklahoma | 1.73 |
| Hawaii | 1.66 | Oregon | 0.46 |
| Idaho | 0.14 | Pennsylvania | 1.22 |
| Illinois | 1.16 | Rhode Island | 1.07 |
| Indiana | 2.10 | South Carolina | 0.92 |
| Iowa | 1.94 | South Dakota | 1.22 |
| Kansas | 1.87 | Tennessee | 1.27 |
| Kentucky | 2.05 | Texas | 1.47 |
| Louisiana | 1.20 | Utah | 2.12 |
| Maine | 0.77 | Vermont | 0.01 |
| Maryland | 1.29 | Virginia | 1.21 |
| Massachusetts | 1.23 | Washington | 0.36 |
| Michigan | 1.41 | West Virginia | 1.99 |
| Minnesota | 1.59 | Wisconsin | 1.71 |
| Mississippi | 1.41 | Wyoming | 2.28 |
| Missouri | 1.88 | U.S. Average | 1.36 |

**Question 26-19:** Use the United States average in Table 26-1 and the 2001 electricity intensity (11,600 kWh per year) to calculate how many pounds of $CO_2$ emissions would be avoided in the United States with a 40 percent improvement in energy efficiency.

According to the EPA, for the year 2003, $CO_2$ emissions from fossil fuel combustion was $1573 \times 10^{12}$ and $1169 \times 10^{12}$ g $CO_2$ equivalents for industrial and residential use, respectively. Total $CO_2$ emissions were $5,562 \times 10^{12}$ g $CO_2$ equivalents.[7]

**Question 26-20:** Calculate the annual savings in emissions for the United States assuming that energy efficiency upgrades could achieve 40 percent improvement in energy efficiency.

**Question 26-21:** Discuss to what extent you think energy efficiency standards should be (a) required by law, (b) encouraged through tax incentives, (c) promoted through government-sponsored education, or (d) entirely market-driven. Justify your answer.

**Question 26-22:** Summarize the major points of this Issue.

**Question 26-23:** Buckminster Fuller said, "The best way to predict the future is to design it." Winston Churchill said, "We shape our buildings; thereafter they shape us." Relate these statements to what you have learned in this issue.

---

[7] Recent data are similar, e.g. for 2012 total $CO_2$ emissions $5,557.6 \times 10^{12}$ g.

# FOR FURTHER THOUGHT

**Question 26-24:**   A 2006 article in the *Myrtle Beach Sun News* described a 13,000 ft² home for a couple in the gated Prince George Community. The house, according to the owners, was the minimum space needed to meet their "wants and needs." Explain whether 13,000 ft² houses can be a part of a sustainable society. If not, discuss ways to discourage people from building such houses or ways to encourage building smaller homes.

**Question 26-25:**   People who live in large houses often justify it with the argument that, since they can afford such a big home, it's nobody's business but their own. How would you respond to this?

**Question 26-26:**   Research the latest trends in sustainable building. What are they?

# Issue 27

## THE THREE R'S: REDUCTION, REUSE, AND RECYCLING

### KEY QUESTIONS

- How much municipal solid waste is recycled?
- What materials have the highest recycling rates?
- What is the difference between pre-consumer and post-consumer waste?
- Is recycling consistent with principles of sustainability?
- How do reuse and reduction differ from recycling?

### RECYCLING

"This is the most exciting thing that's come along for the last 15 or 20 years" in recycling, said Christine Knapp of Citizens for Pennsylvania's Future.[1]

Ms. Knapp was referring to RecycleBank, a nonprofit company started in 2004 by Patrick Fitzgerald and Ron Gonen. Here's how RecycleBank works: Households place recyclables in special bins that have a computer chip (essentially a barcode) embedded in them. Bins are weighed when they are emptied at curbside, and a scan of the barcode records which household the bin belongs to. RecycleBank Dollars, coupons that can be redeemed at various businesses including Starbucks, Home Depot, and Bed Bath & Beyond, or donated to charity, are awarded based on the weight of the recyclables. According to the company, there is virtually no cheating (by adding weight to the bins, for example). RecycleBank charges a fee to municipalities or private waste haulers, who typically offset this cost (or even make a profit) by reduced tipping (dumping) fees. They also receive revenue from some of the recyclable material. The program has tripled recycling rates in Philadelphia neighborhoods and had 3 million participants globally by 2012.

**Question 27-1:** Evaluate whether this is a useful approach to increasing recycling rates. List and evaluate other ways to increase the rate of curbside recycling.

---

[1] Philadelphia residents discover it pays to recycle. *Planet Ark,* www.planetark.org/dailynewsstory.cfm/newsid/35965/newsDate/11-Apr-2006/story.htm; www.recyclebank.com.

Recycling is also catching on at some special events:

- At the 2012 Phoenix Open Golf Tournament, sponsor Waste Management plans to remove trash receptacles and replace them with recycling and composting containers. It expects to divert 90% of potential waste from landfills.
- The Bonaroo Music Festival in Tennessee requires vendors to purchase compostable/recyclable paper products and cups.
- The U.S. Green Building Council's GreenBuild Expo works with a convention center, caterer, and hotels to implement programs to reduce, reuse, and recycle waste at meetings. Some stadiums and arenas are beginning to institute recycling at sporting events.
- In a 2010 survey, 65% of university athletic departments placed a "high" or "very high" emphasis on game-day recycling at athletic events.

**Question 27-2:**    Does your institution or professional team recycle at games or special events? If so, do patrons use recycling bins, or do they merely throw away recyclables?

In 2005, Starbucks, in partnership with Environmental Defense, announced that new disposable paper cups would henceforth contain 10 percent post-consumer waste recycled content, a move that would result in savings of 5 million pounds of virgin tree fiber annually. According to Margaret Papadakis, head of packaging at Starbucks, the process took so long to implement because Starbucks' customers "expect their cups to look a certain way."[2] Starbucks used 4 billion paper cups worldwide in 2011, and proposes to ensure that all of its cups are "recyclable or reusable" by 2015. (We also discuss this topic in Issue 15.)

**Question 27-3:**    In 2004, Starbucks purchased 27,400 tons of 100 percent virgin-bleached cup stock.[3] By how much would landfill content have been reduced by the use of the new cups?

Table 27-1 shows the environmental benefits of using 1.9 billion cups with 10 percent recycled content, according to Environmental Defense.

**TABLE 27-1** ■ Estimated resource savings associated with use of cups with 10% post-consumer waste content by Starbucks.[1]

- 11,300 fewer tons of wood consumed (or about 78,000 trees)
- 58 billion BTUs of energy saved (enough to supply 640 homes for a year)
- 47 million gallons of wastewater avoided (enough to fill 71 Olympic-sized swimming pools)
- 3 million pounds of solid waste prevented (equivalent to 109 fully-loaded garbage trucks)

[1]www.papercalculator.org

[2] www.sijournal.com/breakingnews/1919807.html.
[3] www.starbucks.com/csrnewsletter/winter06/csrEnvironment.asp.

To most Americans, being environmentally sensitive means one thing: recycling. In 2010, according to the EPA, recycling diverted 85 million tons of municipal solid waste (MSW; see Issue 15) from landfills and incinerators. Figure 27-1 shows the composition of MSW, and Figure 27-2 depicts the recycling rates for various materials.

**Question 27-4:** The 85.1 million tons of MSW that were recycled in 2010 represented a recycling rate of 34 percent. How much MSW was not recycled?

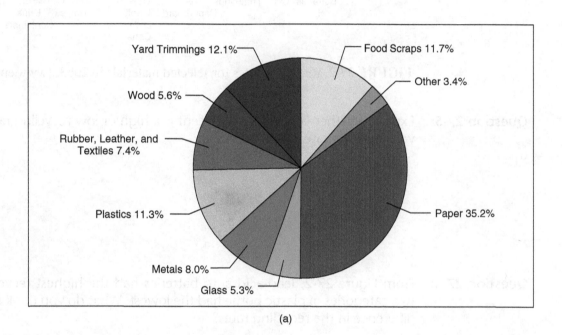

(a)

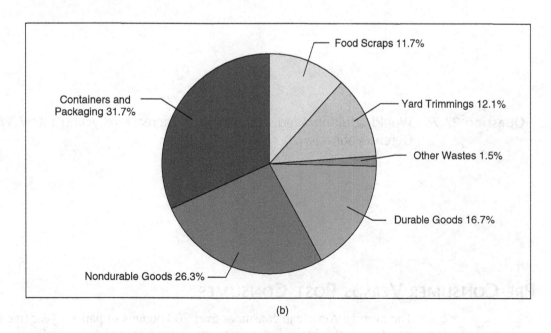

(b)

**FIGURE 27-1** Composition of MSW in 2003. (a) Materials; (b) Products. (www.epa.gov)

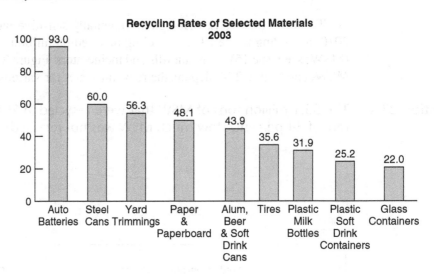

**FIGURE 27-2** Recycling rates for selected materials in 2003. (www.epa.gov)

**Question 27-5:** Explain whether you think 34 percent is a high or low recycling rate. Discuss why you think the rate is not higher.

**Question 27-6:** From Figure 27-2, lead-acid auto batteries had the highest recycling rate, while two categories of plastic bottle had the lowest. What do you think accounts for the difference in the recycling rates?

**Question 27-7:** Would additional legislation help to increase recycling rates? Why or why not? Defend your answer.

## PRE-CONSUMER VERSUS POST-CONSUMER

The average American consumes over 700 pounds of paper a year (the equivalent of nine telephone-pole size trees), nearly 90 percent of which is virgin (i.e., with no recycled content). Unless marked as containing 100 percent post-consumer content (see below), paper that is marked *recycled* also contains virgin fibers, but has some percentage of pre-consumer and/or post-consumer recycled content. Pre-consumer content refers to paper scraps and trimmings produced as a by-product of the paper manufacturing process. Post-consumer waste (PCW) is paper that has been previously used by the end-consumer. The higher the post-consumer waste content, the better the product is from an environmental standpoint. First, high-PCW paper diverts material from landfills, roadsides, or incinerators. In addition to saving trees, high-PCW paper uses significantly less energy and water to manufacture. Finally, many types of high-PCW paper are unbleached or use chlorine-free bleaching processes (which may also be used on some virgin paper products) that produce no dioxin. *Dioxins* are a class of persistent-carcinogenic compounds that are among the most toxic chemicals known (see Issue 17).

**Question 27-8:** The University of Vermont created a paper purchasing policy in 1999 that recommended a minimum of 30 percent PCW content. Does your institution have a recycled paper policy? Check your institution's website, ask an administrative specialist, or call your purchasing department.

In the past, recycled paper did not meet minimum standards for use in high-speed copy machines and printers, but that has now changed.

Book publishers are starting to make the transition to high-PCW paper as well as other innovations. *An Inconvenient Truth,* by Al Gore, was printed on Appleton Green Power 80# Matte paper, which is chlorine-free (and is produced using 100% green power) and contained 30% PCW. The virgin component of Appleton Green Power paper was from sustainably grown trees. According to the *New York Times* (whose products contain on average 27% recycled content), Random House, whose "parent" Bertelsmann buys 3.8 million tonnes of paper annually, is moving toward using 30% recycled paper.[4] Random House's Chief Executive Peter Olsen said that the time has come "to stop making eye-rolling excuses about how burdensome and expensive environmental initiatives are."[5] Efforts by trade publishers and university presses to increase the use of high-PCW paper could decrease greenhouse emissions by over 250,000 tons annually and save 4.9 million trees, 2.1 billion gallons of water, and eliminate 1.3 million tons of solid waste.[6]

An alternative to both virgin and recycled paper is tree-free paper made from fibers such as hemp, sisal, abaca, kenaf, wheat straw, corn, and banana stalks. *Cradle to Cradle,* a book on sustainability by William McDonough, is printed on recycled (and, incredibly, recyclable) plastic.

**Question 27-9:** This textbook was printed on recycled paper. Is it clear what the PCW content is? Write down the titles and publishers of your other textbooks in the space below and what kind of paper (virgin, recycled, high PCW recycled) each was printed on. Ask your instructors if their decisions to adopt textbooks are based to any extent on whether the books contain recycled content, and report your findings here.

[4] Saving the planet, one book at a time, by Rachel Donadio, *New York Times Book Review*, 7-9-06, p. 27; www.bertelsmann.com
[5] Donadio op cit.
[6] Ibid.

# E-WASTE[7]

**Question 27-10:** What makes a computer or cell phone obsolete? Why do you "upgrade" your computer or cell phone? Do the added features enhance the quantity and/or quality of your work? Do they enhance the quality of your life?

In 2010, an estimated 14-20 million computers were thrown "away" in the United States alone. The National Safety Council reports that 250 million computers became "obsolete" by 2009. The "good" news is that legislatures, institutions, and individuals are starting to handle obsolete electronic equipment responsibly. Maine, for example, in 2006 passed a law requiring computer monitor and television manufacturers to pay for their recycling and disposal. Hewlett-Packard has two recycling plants that process 1.5 million pounds of electronics a month. And the manufacture of televisions with cathode-ray tubes (CRTs) is being drastically reduced as televisions with CRTs are replaced with flat-panel televisions. This saves about 8 pounds of lead for each CRT replaced with a flat-panel television.[8]

There is plenty of bad news as well. Elizabeth Royte, reports that many U.S. electronics recyclers simply ship the e-waste to foreign countries with less stringent or absent environmental regulations.[9] After usable materials are salvaged, the remains are dumped in landfills, along roadsides, or in water bodies (see Figures 15-3A, B, and C). All of these dump sites can leach metals like chromium, tin, and barium at toxic concentrations into the ground and nearby streams. A report by the environmental organization Basel Action Network states that 50 to 80 percent of U.S. e-waste was being processed in China, India, Pakistan, and other developing countries under largely unregulated, unhealthy conditions.[10]

**Question 27-11:** If you knew that upgrading to a cell phone or computer meant that your old cell phone or computer would end up in a landfill or contribute to degrading the environment and harming the health of people in developing countries, would you still make the upgrade? Discuss whether you think this is a fair question.

---

[7] We also discuss e-waste in Issue 15.

[8] These and other initiatives are described in the *New York Times* article "Alternatives: Panning e-waste for gold," by Susan Moran (5/17/06).

[9] E-Waste At Large (NYT 1/27/06)

[10] http://www.ban.org/

**Question 27-12:** *New York Times* reporter Laurie J. Flynn quotes the president of Scrap Computers, a recycler in Phoenix, who stated, "There's no such thing as a third-world landfill. If you were to put an old computer on the street, it would be taken apart for the parts."[11] What do you think his point of view was? Re-examine Figures 15-3A, B, and C, then discuss whether you agree or disagree with his statement and his main point.

# REDUCTION AND REUSE

An alternative to recycling is *source reduction*, which is the preferred method of waste management. Source reduction refers to buying less, buying more durable products, buying products with reduced packaging, and buying products designed to be reused.

**Question 27-13:** List consumer products that are designed to be useful for a short period (planned obsolescence), that are overpackaged, or that are not designed for reuse (after original use).

**Question 27-14:** If source reduction is the preferred method of waste management, why do you suppose the focus of most waste management programs is recycling?

# THE FUTURE OF PLASTIC

Most plastic is a petrochemical product, some of which is recycled. One of the problems with plastic recycling (with the possible exception of William McDonough's book described above) is that it does not "close the loop." That is, unlike paper and aluminum, post-consumer plastic is not remanufactured into the same product that was recycled. For example, recycled soda bottles are not remade into soda bottles, but instead are manufactured into products that can use a lower grade of plastic (e.g., carpet, toys, or fleece). According to Eureka Recycling, one of the biggest non-profit recyclers in the United States, recycling plastic only delays its disposal. "The final destination for all plastic is either a landfill, where it doesn't decompose, or an incinerator, where it releases harmful chemicals when burned."[12]

---

[11] Poor nations are littered with old PC's. *New York Times.* 10/24/05.
[12] www.articleworld.org/index.php/Plastic_recycling.

**Question 27-15:**   Major uses of plastic include disposable 2-liter bottles for sugary drink products and bottled water. Discuss the advantages of using plastic as containers for these products. Evaluate whether these are sustainable uses of plastic. List and discuss alternatives to these containers.

**Question 27-16:**   Is shipping truckloads of bottled water thousands of miles using fossil fuels a sustainable practice? Could it be made sustainable? How?

**Question 27-17:**   Considering that plastic production and recycling are fraught with environmental and environmental health issues, should we continue to use plastics as widely as we do? If not, discuss ways that plastic use could be reduced.

Another alternative to recycling is reuse, which includes personal choices like using travel mugs instead of disposable cups, and buying second-hand clothing and furniture.

**Question 27-18:**   List additional examples of source reduction and reuse that you could adopt.

**Question 27-19:**   Summarize the major points of this Issue.

# FOR FURTHER THOUGHT

## Focus: Aluminum Mining and Refining

Aluminum packaging is a mainstay of the beverage industry. Let's examine the environmental cost as well as the subsidies involved in the manufacture of the all-aluminum can.

Aluminum (Al) is one of the most common elements in the Earth's crust, but ores of aluminum are rare. The most common ore of aluminum is *bauxite*. Bauxite was first discovered in the French district of Les Baux in 1821, from which it took its name. Bauxite is found in extensive deposits in many countries, but mainly in the tropics and subtropics. Even though bauxite is a rich ore, the aluminum metal is tightly locked into the mineral structure and is not easily extracted.

First the bauxite must be crushed; then it is processed into a white powder called alumina ($Al_2O_3$). During processing, four tonnes of bauxite ore yield two tonnes of alumina and two tonnes of waste.[13] The two tonnes of alumina further reduce to one tonne of Al metal.

**Question 27-20:** How much waste is produced per tonne of Al refined?

The alumina is then dissolved in molten cryolite ($Na_3AlF_6$) to make the solution conduct electricity. An electric current separates the Al from the solution and molten aluminum metal sinks to the bottom of the vat, where it is drawn off. A by-product of the reaction, HF, is extremely toxic and is released as a gas. $CO_2$, a greenhouse gas, may also be emitted in the process.

## Aluminum Cans

Cans are made from thin sheets of refined aluminum. The efficiency of can manufacture has been increasing. In 1972, 21.75 cans were made from a pound of aluminum. In 2008, that number had increased to 34.2.[14]

Aluminum melts at 660°C, so it's easy to recycle. It takes only 5 percent of the energy to make new cans from old cans (scrap) compared to what it takes to make new cans from bauxite, and of course, a great deal of waste is avoided as well (see above). For example, for each pound of aluminum produced by recycling scrap, 3 pounds of waste are not generated.

Consider these data:

| Year | Recycling Rate (%) |
|------|--------------------|
| 1973 | 15.2 |
| 1974 | 17.5 |
| 1975 | 26.9 |
| 1976 | 24.9 |
| 1977 | 26.4 |
| 1978 | 27.4 |
| 1979 | 25.7 |

**Question 27-21:** Plot these data on the axes below, then project the recycling rate for 2010. (There are a variety of ways to do this: You can either extrapolate from your graph or calculate the most recent growth rate and use $N = N_0 \times e^{rt}$.)

---

[13] www.alcan.com.
[14] Aluminum Association. www.aluminum.org.

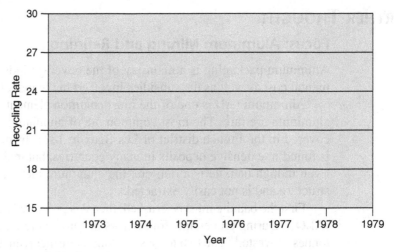

Here are the rest of the data:

| Year | Recycling Rate (%) |
| --- | --- |
| 1980 | 37.3 |
| 1981 | 53.2 |
| 1982 | 55.5 |
| 1983 | 52.9 |
| 1984 | 52.8 |
| 1985 | 51.0 |
| 1986 | 48.7 |
| 1987 | 50.6 |
| 1988 | 54.6 |
| 1989 | 60.8 |
| 1990 | 63.6 |
| 1991 | 62.4 |
| 1992 | 67.9 |
| 1993 | 63.1 |
| 1994 | 65.4 |
| 1995 | 62.2 |
| 1996 | 63.5 |
| 1997 | 66.5 |
| 1998 | 62.8 |
| 1999 | 62.5 |
| 2000 | 54.5 |
| 2001 | 55.4 |
| 2002 | 48.4 |
| 2003 | 50.0 |
| 2004 | 51.2 |
| 2005 | 52.0 |
| 2006 | 51.6 |
| 2007 | 53.8 |
| 2008 | 54.2 |
| 2009 | 57.4 |
| 2010 | 58.1 |

**Question 27-22:**   Plot the recycling rates from 1972 to 2010 on the axes below. Did the recycling rate you projected from the 1972–79 data agree with the actual one? Did this reinforce or challenge your faith in the accuracy of projections? Why?

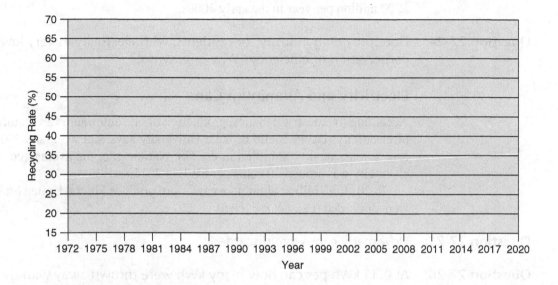

**Question 27-23:**   On your graph, use a dashed line to project a recycling rate for 2020. How comfortable are you with your projection? Explain.

## Subsidies and Aluminum Production

The electricity to extract the aluminum is only part of the cost to produce aluminum cans. Other costs include the extraction and shipping of the ore, the cost to restore the surface environment after mining is completed (or if restoration is not carried out, this cost is simply "dumped" onto the local environment and residents), the energy to ship the ore or concentrate to the refinery, and the cost to transport and distribute the cans. Economists define *externalities* as a cost that is not included in the price users are charged for a commodity. Subsidies are a type of externality.

Subsidies tend to distort markets, since subsidies encourage use of a commodity in excess of the level that the commodity would be used if all the costs to produce the commodity were included in the price. See Issue 16 for more discussion of mining subsidies. For example, the Australia Institute reports that the Australian government awarded mining companies subsidies of around $4 billion (Au) per year.[15]

## Aluminum Refining and Electricity Costs

Refining aluminum from ore is extremely energy intensive, so most aluminum refiners either operate their own power dams, or concentrate in areas where electric rates are low. Such an area is the Pacific Northwest, where a federal agency established in 1937 by the Roosevelt administration, the Bonneville Power Authority (BPA), operates power dams on the Snake and Columbia Rivers. Under legislation that created the BPA, the agency must first serve public power agencies such as municipal utilities, as well as rural cooperatives and half a dozen government agencies. Some of what is left must go to investor-owned utilities. What is left over has been customarily sold to private industries, such

---

15   The Australia Institute, https://www.tai.org.au/index.php?q=node%2F19&pubid=986&act=display.

as the aluminum companies. Where there was once plenty of leftover power, population growth in the Northwest and a booming high-tech economy has led to tight power supplies. By 2000, Bonneville was forced to go into the market and purchase about 1,000 megawatts of capacity. BPA provided subsidies to aluminum refiners (in the form of cheap power) at $200 million per year in the early 2000s.

**Question 27-24:** Does providing industry, or residents, with electricity at very low rates encourage conservation or efficiency? Why or why not?

### Electricity and Aluminum Cans

According to industry statistics, production of one aluminum can requires about 0.35 kWh of electricity. David Biello of Yale University says that we make 300 billion aluminum cans a year, about 100 billion in the US. Nationwide, the average cost to *individuals* for electricity was around 10 cents per kWh in 2012.

In 2011, 98 billion aluminum cans were produced in the United States, and 42 percent were discarded (i.e., not recycled).

**Question 27-25:** How many cans were discarded?

**Question 27-26:** At 0.35 kWh per can how many kWh were thrown away with the aluminum?

The Sierra Club and the Competitive Enterprise Institute, among others, favor eliminating subsidies for electricity.[16] Many scientists urge increasing water flow around the BPA's dams to improve the rivers' environments for salmon and steelhead trout, which before dam construction supported a thriving fishing industry on these rivers. Some environmentalists favor dismantling many of the dams altogether.

Even though aluminum producers strongly support recycling and have underwritten thousands of recycling centers around the country (spending hundreds of millions of dollars in the process), the responsibility to deal with solid waste rests with local government, not with the producers of packaging. Can manufacturers point out that facilities to produce aluminum cans from scrap take half the time to build and cost one-tenth the amount of a mine and refinery.

**Question 27-27:** Is it fair to say that electricity used to produce thrown-away cans could be used for more productive activities? Discuss and explain your reasoning.

**Question 27-28:** Should producers of packaging be responsible for their product? Why or why not? Justify your conclusion.

**Question 27-29:** Do you think the destruction of fishing industries from rivers dammed to produce cheap hydroelectric power is a reasonable price to pay for artificially cheap aluminum cans that contain drinks with little or no nutritional value? Was this a biased question? Explain.

---

[16] www.cei.org

# A SUSTAINABLE DIET

## KEY QUESTIONS

- What is myplate.gov?
- What constitutes a sustainable diet?
- Is it practical to have a sustainable diet?
- What are the medical, social, and environmental consequences of the standard American diet?
- What fundamental changes are necessary if our diet is to become sustainable?
- What is the relation of industrial hog farms to sustainability?

## MYPLATE.GOV

In 2010, the U.S. Department of Agriculture radically revised its Dietary Guidelines for Americans. It began, for the first time, to encourage Americans to eat a plant-focused, and therefore more sustainable and healthy, diet. It's website www.ChooseMyPlate.gov offers dietary goals that could help change the dangerous and unsustainable trends of eating in the United States, which have led to epidemics of Type 2 diabetes, obesity, and cardiovascular disorders, in turn resulting in hundreds of billions in unnecessary health care costs. Here are the Department's recommendations:[1]

- Avoid oversized portions
- Eat fish twice a week, watch out for mercury in fish, and eat plant protein foods more often
- Make at least half your diet fruits and vegetables
- Make at least half your grains whole grains
- If you drink milk, switch to low fat or nonfat milk
- If you eat processed foods, choose low-sodium processed foods
- Drink water instead of sugary drinks

What the recommendations do not do is encourage Americans to educate themselves on the meaning of a sustainable diet. Recall that sustainability refers to meeting today's resource needs without limiting the ability of future generations to meet theirs. Sustainability embraces the *triple bottom line* of people, prosperity, and planet and includes the goals of social and economic justice. Throughout this book, we have examined barriers to sustainability and ways in which people, communities, and institutions can move toward

---

[1] www.ChooseMyPlate.gov.

sustainability. We gave examples of constructing a LEED-certified building (Issue 26), instituting smart growth policies (Issue 24), transforming a university (Issue 29), or restoring Southeast long-leaf pine ecosystems (Issue 30).

The American diet, however, seems to be moving in the opposite direction. Our consumption of meat is among the world's highest (over 170 pounds per person annually),[2] and this is encouraging industrial streamlining of the process by which animals are raised for slaughter. This, in turn, has led to unethical and inhumane living conditions for cows, calves, pigs, chickens, and other livestock, and dangerous working conditions for employees.

In addition to being meat-centered, the conventional American diet is high in "bad" fats and simple, refined carbohydrates obtained from snack foods, sugary soft drinks, and fast foods and, perhaps most importantly, lacking in fiber from fresh fruits and vegetables. The result is the epidemic of coronary artery disease, type 2 diabetes, and obesity that we referred to earlier.[3]

In 2006, author Jeremy Weinstein published *The Ethical Gourmet: How to Enjoy Great Food That is Humanely Raised, Sustainable, Nonendangered, and That Replenishes the Earth.*[4] The book contained "The Ethical Gourmet's Ten Commandments," which focused not only on a healthful diet, but also on a sustainable one.

Here are *The Ethical Gourmet's* Ten Commandments:

1. Turn the menu upside down, so meat is an accompaniment, not the centerpiece.
2. Choose local foods over imported organic, organic foods over conventional.
3. Support responsible development of genetically modified foodstuffs to reduce pesticide abuse and deforestation.
4. Choose fish from closed aquaculture systems, farmed shellfish, and fast-reproducing wild fish from managed fisheries.
5. Drink tap, not bottled, water.
6. Dine at restaurants that serve a moral menu.
7. Look for humane, sustainable, and ethical logos on packaged food, such as Certified Humane Raised and Handled, Rainforest Alliance, Fair Trade, and the Marine Stewardship Council. Carry the Blue Ocean Institute seafood guide in your wallet.
8. Buy from companies with responsible environmental policies and invest in socially responsible funds.
9. Vote for candidates who are committed to ending agricultural subsidies and a laissez-faire approach to regulation in the meat and dairy industries. Lobby those who aren't.
10. Cook wholesome food at home more often.[5]

**Question 28-1:** In what ways are the USDA recommendations similar to Weinstein's "Ten Commandments?" In what ways do they differ?

---

[2] U.S. Department of Agriculture; U.S. Bureau of the Census.
[3] Centers for Disease Control and Prevention (www.cdc.gov/nccdphp/dnpa/obesity/trend/index.htm).
[4] Broadway Books.
[5] Reprinted from *The Ethical Gourmet* (2006), Random House, Inc.

Many organizations that are (or should be) aware of the environmental impacts of industrial meat production and the health consequences of the conventional American diet continue to feature unsustainable menus at gatherings and special events. A regional ecological society hosted a "pig-pickin" at a conference not long after heavy rains associated with a tropical storm caused a hog-waste lagoon to breach, killing thousands of fish in a nearby estuary being studied by members of the same organization. Sugary soft drinks served in disposable cups are mainstays at informal gatherings of environmental groups. And fast food franchises are commonplace on college campuses (where obesity rates are plainly evident) and, perhaps most surprisingly, hospitals.[6]

# THE SUSTAINABLE DIET

**Question 28-2:** According to the Sustainable Table website, "Sustainable agriculture is a way of raising food that is healthy for consumers and animals, does not harm the environment, is humane for workers and animals, provides a fair wage to the farmer, and supports and enhances rural communities."[7] Go to their website and describe the five principles of sustainable agriculture.

**Question 28-3:** Record your own diet for the previous 48 hours. In general, describe what foods you eat regularly, where and how frequently you eat out, etc. How sustainable and healthy do you perceive your diet to be?

## Meat Consumption and A Sustainable Diet

In 2012 the USDA estimated world meat production to be 249 million tonnes, an all-time high.[8] See Table 28-1. Because of pork's dominance in China, it ranks first globally. In the United States, 92 billion pounds of red meat and poultry were produced in 2010, with beef first and poultry a close second. [9]

**TABLE 28-1** ■ World Meat Production, 2012 (tonnes) (Source: USDA)

| Beef and Veal | 57,000,000 |
|---|---|
| Pork | 104,300,000 |
| Fowl (chicken and turkey) | 87,500,000 |

---

[6] Weil, A. Surgery with a side of fries. *New York Times*, July 6, 2006.
[7] www.sustainabletable.org/intro/whatis/.
[8] USDA, op cit.
[9] Statistical Abstracts of the United States, 2010. www.census.gov.

**Question 28-4:**    What is the per capita global meat production for 2012 (mid-2012 global population = $7 \times 10^9$)?

Many experts have now concluded that the centerpiece of any program of sustainable eating minimizes or even eliminates meat. Figure 28-1 gives a few reasons why.

To meet global and domestic demand for meat (often stimulated and maintained by trade organizations), U.S. farmers have adopted highly efficient industrial processes in which animals are grown quickly to a marketable size, often packed together under what most would consider inhumane conditions (Figure 28-2).

According to Caitlin Winans, former director of the Coalition to Reform Animal Factories,

> From an animal welfare perspective, most chicken meat and eggs are among the worst food choices you can make. The way most chickens are raised in confined animal feeding operations

---

## The Ins and Outs of Meat Production

**INPUTS**

**Feed**
• A calorie of beef, pork, or poultry needs 11–17 calories of feed.
• 80 percent of soybean harvest is eaten by animals, not people.

**Water**
• Producing 8 ounces of beef can require up to 50,000 liters of water.

**Additives**
• Cows, pigs, and chickens get 70 percent of all antimicrobial drugs in the United States.

**Fossil Fuels**
• According to the UN's Food and Agricultural Organization (FAO)* "the ranching and slaughter of cows and other animals generates an estimated 18 percent of total human-induced greenhouse gas emissions globally."

**OUTPUTS**

**Manure**
• Manure from intensive pig operations stored in "lagoons" can pollute groundwater and surface water.

**Methane**
• Belching, flatulent livestock emit about 16 percent of the world's annual production of methane, a powerful greenhouse gas.

**Disease**
• Eating animal products high in saturated fat and cholesterol is linked to cancer, heart disease, and other chronic illnesses.
• Factory farm conditions can spread *E. coli*, *Salmonella*, and other food-borne pathogens.
• Creutzfeldt-Jakob disease, the human variant of mad-cow disease, has killed at least 150 people.

**Dead Zones**
• The Dead Zone in the Gulf of Mexico, one of many worldwide, has averaged 17,000 square km over the past decade, as large as Lake Ontario. Its cause is mainly nutrient runoff from agriculture in the Mississippi Basin. As much as eighty percent of the grains grown in the United States are fed to animals.

---

\* Livestock's Long Shadow, http://www.fao.org/docrep/010/a0701e/a0701e00.HTM
\*\* www.noaa.gov; http://www.hulfhypoxia.net/Overview

**FIGURE 28-1**    The environmental cost of consuming meat. (Nierenberg, D. 2005. *Worldwatch Paper #171: Happier Meals: Rethinking the Global Meat Industry*. (www.worldwatch.org)

**FIGURE 28-2** Egg-laying hens confined on a typical egg farm in the United States. Compassion Over Killing. (www.cok.net).

(CAFOs) is even more inhumane than how other farm animals are raised. Additionally, the USDA doesn't even include poultry in the Humane Methods of Slaughter Act, leaving literally billions of farm animals a year to die in an unnecessarily cruel way. Currently more than 90 school campuses have made steps to improve the lives of egg-laying hens by enacting policies to eliminate or greatly reduce their use of eggs from caged hens.[10]

**Question 28-5:** Are you aware of industrial livestock practices? If so, discuss whether this awareness has any impact on the type and quantity of meat you eat. If not, discuss whether awareness of these practices would have an impact on the type and quantity of meat that you eat.

**Question 28-6:** Select any two of Weinstein's "commandments" (except #1), and evaluate the rationale behind them.

---

[10] Personal Communication.

**Question 28-7:**    Discuss whether you support these commandments. Would it be difficult to comply with these commandments? Why or why not?

# COASTAL POPULATION GROWTH (OF HOGS)

NEWS ITEM: On August 8, 1995, in Brunswick County, North Carolina, a "lagoon" containing waste from 6,400 head of swine ruptured. Approximately 7.6 million L of manure was discharged into nearby streams, which emptied into the Cape Fear River. The Cape Fear River flows into the Cape Fear estuary.[11]

What impact does a spill of this magnitude have on aquatic ecosystems? How frequently do such spills occur? Who should be held responsible?

Coastal North Carolina has quietly become one of America's most important centers for hog raising and processing. By 2005, North Carolina's 16 million hogs outnumbered its 9 million people. The hogs were concentrated in the coastal plain east of Interstate 95, and those hogs were producing waste at the rate of 19 million tonnes per year.

**Question 28-8:**    On average, hogs produce 6.4 pounds of waste per day, while humans produce 2.3 pounds. Contrast the total waste produced by North Carolina's human and hog populations per day for 2005.

Hog factory farms confine pigs in stalls with eight square feet of floor space per animal in metal buildings. *60 Minutes* correspondent Morley Safer reported that hogs ". . . see no sun in their limited lives, with no hay to lie on, no mud to roll in. The sows live in tiny cages, so narrow they cannot even turn around."[12] The floors are slatted metal, through which feces and urine fall. Hog waste is officially regulated by government agencies, but inspections are rarely carried out because of insufficient staff. There are two legal ways of disposing of the waste: (1) spraying it onto fields and (2) washing it into lined pits, which are euphemistically called *lagoons.* In the first case, the nutrients in the waste would ideally be utilized by surface plants and neutralized, and in the second, the waste would be stored until it composted, at which time it would be removed and used as fertilizer, digested to produce methane (natural gas), or buried.

The Chesapeake Bay Commission reports that applying waste to fields allows much of the soluble nitrates and phosphates to seep into groundwater or flow into surface creeks, since it is usually applied at rates that vastly exceed the ability of the plants to take up the nutrients.[13] Thus, high levels of nitrates, which can cause death in extreme cases if ingested by small infants, can build up in well water used by people who live near the "sprayfields."

---

[11]  Mallin, M., et al. 1997. Comparative effects of poultry and swine waste lagoon spills on the quality of receiving streamwaters. *J. Env. Qual. 26:* 1622–1631.

[12]  Morley Safer, "60 Minutes," 9/19/97

[13]  www.chesapeake.org/pubs/wrkshops/P-report.pdf.

Also, sewage lagoons are *legally permitted to leak.*[14] At the maximum allowable rate of 0.036 inches per day, a three-acre lagoon can leak more than a million gallons of waste into groundwater annually. And since there are hundreds of such lagoons in the coastal region of North Carolina, the problem is a regional one.

Spills from sewage lagoons are not uncommon. Ten million fish were killed and over 350,000 acres of salt marshes were closed to shellfishing in 1995 after 25 million gallons of animal waste spilled from an eight-acre waste-pit into North Carolina's New River.[15] In September of 1999, the region was hit by two tropical storms, and the last, Floyd, caused billions of dollars in damage, thankfully killing few people but tens of thousands of hogs and millions of turkeys and chickens, most of which are raised using industrial agricultural methods. Marine Corps helicopters airlifted thousands of hog carcasses, weighing upwards of 250 pounds each, for destruction in portable incinerators furnished by the federal government.

Dozens of sewage treatment plants were washed out or decommissioned, and pits holding hog waste were washed away while towns such as Princeville and Tarboro were under water for days. Much of the damage was not measurable for months and may not yet be apparent. Already scientists have described what they refer to as the "Cape Fear Plume," a mass of polluted water covering hundreds of square kilometers to a depth of 10 meters.

In addition to nitrogen and phosphorous pollution and groundwater contamination, other environmental consequences of industrial hog farming include air pollution from ammonia, odor problems, the presence of pathogens, development of antibiotic resistance, disposal of heavy metals from waste lagoons, and generation of greenhouse gases from manure at animal feedlots.[16]

The stench problem cannot be understated. A June 3, 2004, report on PBS's *The News Hour* quoted two residents whose church is half a mile from a large hog farm. According to Evelyn Powell, "The kids say it stinks out there. They get nauseous, they get sick . . . , and they are unable to go outside and play and enjoy our quality of life at the day care."[17]

**Question 28-9:** How would these impacts described by residents show up in an economic analysis of the impact of industrial-scale hog raising? Is it possible to quantify the suffering these residents have expressed? Explain your answer.

**Question 28-10:** Do the impacts described by residents constitute a subsidy for industrial hog production? Explain.

[14] Environmental Defense. 2000. Environmental Impacts of Hog Factories in North Carolina. http://www.environmentaldefense.org/documents/2537_Hogwatch_EnviroImpacts.pdf#search='hog%20lagoon%20leak%20rate.'
[15] Marks, Robbin. *Cesspools of shame: How factory farm lagoons and sprayfields threaten environmental and public health.* Natural Resource Defense Council and the Clean Water Network. July 2001.
[16] Wright, Christian L. Many Little Piggies, Handled With Care. *New York Times,* May 17, 2006, p. G10.
[17] www.pbs.org/newshour/bb/science/jan-june04/pigs_6-3.html.

Yet another concern is injuries to workers in hog processing plants. *New York Times* columnist Bob Herbert focused one of his opinion columns on the largest pork-processing facility in the world, the Smithfield Packing Company (Tar Heel, North Carolina). Herbert wrote,

> It's a case study in both the butchering of hogs (some 32,000 are slaughtered there each day) and the systematic exploitation of vulnerable workers. More than 5,500 men and women work at Smithfield, most of them Latino or black, and nearly all of them undereducated and poor.[18]

He quotes worker Edward Morrison:

> . . . You have to work fast because that machine is shooting those hogs out at you constantly. You can end up with all this blood dripping down on you, all these feces and stuff just hanging off of you. It's a terrible environment.[19]

**Question 28-11:** Refer to the definitions of sustainability in this Issue's introduction, especially the concept of the triple bottom line. Discuss whether North Carolina's hog farming industry is sustainable.

**Question 28-12:** Do you believe consumers are aware of the environmental, ethical, and human rights dimensions of the food they eat? Are you? Discuss.

Sustainable, or near-sustainable, hog farming does exist.[20] The key here is *small*—small hogs, small herd size, small farm. Wastes are recycled on site. Use of antibiotics is reduced or eliminated. Food for the hogs is grown on the farm. And the animals are given sufficient space to roam (Figure 28-3). Once grown to marketable size, the hogs are sold through alternative channels that do not exploit workers. The only caveat: Economists doubt that small-scale operations could meet large-scale demands.

**Question 28-13:** Summarize the major points of this Issue.

---

[18] Where the Hogs Come First. *New York Times,* June 15, 2006, p. A 17. 21Ibid.

[19] Ibid.

[20] https://attra.ncat.org/.

**FIGURE 28-3** Free range hogs. (Courtesy of Travel Ink/Getty Images USA, Inc.)

## FOR FURTHER THOUGHT

Caitlin Winans, then with the Coalition to Reform Animal Factories[21], adds,

> It's important to think about the way we talk about environmental issues. The Coalition to Reform Animal Factories will not use the word *farm* when talking about confined animal feeding operations (CAFOs) because it conjures up images of red barns, rolling hills and happy animals. In contrast, CAFOs are usually windowless, crowded, and polluted factories where animals are treated like production units rather than animals. They should be referred to as *animal factories*. Another example is the CAFO industry's use of the word *lagoon*. We refer to those [more] accurately as *cesspools*.[22]

**Question 28-14:** Discuss whether you agree or disagree about the importance of language to environmental issues. Which terms (*farms* or *animal factories*; *lagoons* or *cesspools*) do you think most accurately reflects the reality of the settings? List other examples of language either exaggerating or understating the severity of any environmental impact.

**Question 28-15:** For 2009, AdWeek reported that "Beverage Digest estimates that per-capita American consumption of carbonated soft drinks in 2009 fell to 736 eight-ounce servings, down from 760 servings in 2008."[23]

How much is that per day per capita? There are about 100 calories in an 8-ounce "serving." How many calories does the average American consume in one day from sodas?

**Question 28-16:** Discuss whether soft drink consumption can be a part of a sustainable diet. Under new rules, in New York City soft drinks in cups larger than 16 ounces cannot be sold after February 2016. Do you support this measure? Why or why not?

---

[21] Currently with the Pew Environment Group.
[22] Personal communication, op cit.
[23] http://www.adweek.com/news.

**Question 28-17:** According to the Christian Science Monitor, 80 percent of egg cartons are imprinted with the logo "Animal Care Certified."[24] This designation means that the United Egg Producers, the egg industry's trade group, has certified egg producers according to internal guidelines.

    a. What do you think the typical consumer interprets "Animal Care Certified" to mean?

    b. Says Paul Shapiro, executive director of Compassion Over Killing, "The logo is a scam. It conveys the message that the birds are humanely cared for, which couldn't be further from the truth."[25] Mitch Head, United Egg Producers spokesman, responded "There's nothing dishonest about saying our eggs are 'animal care certified.' Consumers can be assured that eggs produced under animal care guidelines have been carefully checked out according to the highest scientific standards."[26]

What do you conclude each responder means? What is the point of view of each? Are they discussing the same concept? Discuss who, if anyone, should be responsible for certifying eggs.

**Question 28-17:** Access the website of Smithfield Foods (www.smithfieldfoods.com/Enviro/). What is their side of the "factory farm" controversy?

---

[24] Wolcott, Jennifer. Cage-free eggs not all they're cracker up to be. *Christian Science Monitor*, October 27, 2004.

[25] Ibid.

[26] Ibid.

# Issue 29

# THE SUSTAINABLE CAMPUS

## KEY QUESTIONS

- What is higher education doing to promote sustainability?
- Who are the leaders among institutions of higher learning?
- What constitutes a sustainable campus?
- Does your institution practice sustainability?
- What can you do to promote sustainability on your campus?

## WHY HIGHER EDUCATION?

Our global environment is in serious trouble. In previous Issues, you have examined the key source of these problems: too many people using too many resources, particularly fossil fuels. Virtually every indicator of planetary health, from the state of our forests to the health of aquatic ecosystems, is declining, and some environmental indicators are in grave condition (e.g., many marine fisheries).

Who is responsible for this state of affairs? As residents of the United States, we shoulder a disproportionate share of the responsibility. We represent less than 5 percent of the world's population but use about 20 percent of its resources and produce about the same percentage of climate-altering greenhouse gases. But Oberlin College professor David Orr blames college-educated people and the institutions that educate them.[1] Why single out higher education when obtaining a college degree is such a desirable achievement? Here's why, according to Orr: College graduates hold almost all positions of authority and power and earn the most money. They therefore disproportionately make the decisions that set the planet's (and our nation's, state's, county's, etc.) course by how they govern, how they lead corporations, what they live in, what they drive, what they eat, and what they buy.

Orr and others blame the modern university for producing graduates who may be competent professionals but who do not know how to live a reasonable, sustainable life. The discipline of economics is particularly guilty, because it has developed "accounting systems [which] do not subtract the costs of biotic impoverishment, soil erosion, poisons in our air and water, and resource depletion from gross national product."[2] But all academic disciplines share a portion of the blame. Even in the face of incontrovertible evidence that our planet's life support system is failing, academicians continue to promote an obsolete curriculum with artificial disciplinary lines and core curricula that seem to be designed to give a piece of the curricular pie to each of the disciplines rather than respond to the planetary emergency that threatens our prosperity and survival on Earth.

---

[1] Orr, D. 1994. *What is education for?* Island Press, Washington. p. 9.
[2] Ibid.

In the words of Orr: "The planetary emergency unfolding around us . . . is . . . first and foremost, a crisis of thought, values, perspectives, ideas and judgment. In other words, it is a crisis of mind, which makes it a crisis for those institutions which purport to improve minds. This is a crisis of education, not one in education."[3] Anthony Cortese, president of Second Nature and former director of the Center for Environmental Management at Tufts University, has stated, "Rather than being isolated in its own academic discipline, education about the environment . . . must become an integral part of the normal teaching in all disciplines."[4]

However, a major sustainability movement is afoot in higher education. In 1990, *University Leaders for a Sustainable Future*[5] produced the Talloires (pronounced *tal-WHAR*) Declaration (Figure 29-1), the first affirmation of sustainability principles in higher education. This document has been signed by the presidents of more than 400 colleges and universities globally. In addition to promoting the practice of institutional sustainability, this initiative has led to the establishment of programs in sustainability or sustainable development across the disciplines in higher education.

---

### Association of University Leaders for a Sustainable Future

#### The Talloires Declaration
#### 10 Point Action Plan

We, the presidents, rectors, and vice chancellors of universities from all regions of the world are deeply concerned about the unprecedented scale and speed of environmental pollution and degradation, and the depletion of natural resources.

Local, regional, and global air and water pollution; accumulation and distribution of toxic wastes; destruction and depletion of forests, soil, and water; depletion of the ozone layer and emission of "green house" gases threaten the survival of humans and thousands of other living species, the integrity of the earth and its biodiversity, the security of nations, and the heritage of future generations. These environmental changes are caused by inequitable and unsustainable production and consumption patterns that aggravate poverty in many regions of the world.

We believe that urgent actions are needed to address these fundamental problems and reverse the trends. Stabilization of human population, adoption of environmentally sound industrial and agricultural technologies, reforestation, and ecological restoration are crucial elements in creating an equitable and sustainable future for all humankind in harmony with nature.

Universities have a major role in the education, research, policy formation, and information exchange necessary to make these goals possible. Thus, university leaders must initiate and support mobilization of internal and external resources so that their institutions respond to this urgent challenge.

We, therefore, agree to take the following actions:

*1) Increase Awareness of Environmentally Sustainable Development*
Use every opportunity to raise public, government, industry, foundation, and university awareness by openly addressing the urgent need to move toward an environmentally sustainable future.

*2) Create an Institutional Culture of Sustainability*
Encourage all universities to engage in education, research, policy formation, and information exchange on population, environment, and development to move toward global sustainability.

*3) Educate for Environmentally Responsible Citizenship*
Establish programs to produce expertise in environmental management, sustainable economic development, population, and related fields to ensure that all university graduates are environmentally literate and have the awareness and understanding to be ecologically responsible citizens.

*4) Foster Environmental Literacy For All*
Create programs to develop the capability of university faculty to teach environmental literacy to all undergraduate, graduate, and professional students.

**FIGURE 29-1**  The Talloires Declaration. (Reprinted from *Association of University Leaders for Sustainable Future* (1994), by permission of University Leaders for a Sustainable Future.)

---

[3] Orr, D. 1994 "The Greening of Education" www.schumacher.org.uk/lec_arch_10-94_bristol.htm.
[4] The Essex Report: Workshop on the principles of sustainability in higher education. www.secondnature.org/history/writings/article/essex_report.htm.
[5] www.ulsf.org.

*5) Practice Institutional Ecology*

Set an example of environmental responsibility by establishing institutional ecology policies and practices of resource conservation, recycling, waste reduction, and environmentally sound operations.

*6) Involve All Stakeholders*

Encourage involvement of government, foundations, and industry in supporting interdisciplinary research, education, policy formation, and information exchange in environmentally sustainable development. Expand work with community and nongovernmental organizations to assist in finding solutions to environmental problems.

*7) Collaborate for Interdisciplinary Approaches*

Convene university faculty and administrators with environmental practitioners to develop interdisciplinary approaches to curricula, research initiatives, operations, and outreach activities that support an environmentally sustainable future.

*8) Enhance Capacity of Primary and Secondary Schools*

Establish partnerships with primary and secondary schools to help develop the capacity for interdisciplinary teaching about population, environment, and sustainable development.

*9) Broaden Service and Outreach Nationally and Internationally*

Work with national and international organizations to promote a worldwide university effort toward a sustainable future.

*10) Maintain the Movement*

Establish a Secretariat and a steering committee to continue this momentum, and to inform and support each other's efforts in carrying out this declaration.

1994 Updated Version

**FIGURE 29-1** *(continued)*

**Question 29-1:** By 2012, 440 institutions worldwide had signed the Declaration. Has your institution signed the Talloires Declaration?[6] If so, what tangible steps has it taken to implement principles of sustainability?

Two landmark events took place in 2005: the Association for the Advancement of Sustainability in Higher Education was launched and the United Nations Decade of Education for Sustainable Development was begun.

Earlier, the National Wildlife Federation's Campus Ecology Program was founded to promote sustainability on college campuses. National student groups, including the Student Environmental Action Coalition and Sierra Club chapters, as well as local student clubs (like Coastal Carolina University's Students for Environmental Action) have also begun to transform institutions. Tufts University started an extracurricular activity—the EcoReps program—to promote student participation in making campuses sustainable, and this model has been adopted at numerous colleges and universities.

A welcome development in higher education has been the opening of sustainability offices and the hiring of sustainability coordinators on campuses across the United States. In 2006, the Coastal Carolina University Campus and Community Sustainability Initiative was launched, with one of us (DCA) as its original director. The Initiative's goals mirrored others around the country: (1) transform the campus into a sustainable university, (2) promote the inclusion of sustainability across the curriculum, and (3) work with the surrounding community to transform it.

---

[6] If you do not know, ask an employee of your institution, check its website, or go to www.ulsf.org and follow links to the list of signatory institutions.

Here is a sample of some of the changes occurring in higher education:[7]

■ Since 2005, all construction at Emory University in Atlanta, Georgia, has been LEED-certified. LEED (Leadership in Education and Environmental Design, see Issue 26) buildings are energy and water-efficient, have improved indoor air quality, use building materials and procedures having low environmental impact, and preserve the natural landscape.

■ The University of South Carolina built the country's largest LEED-certified dorm complex (the West Quad dorm). It features light wells for daylighting, waterless urinals, non-degassing carpets, low volatile organic compound (VOC) paint, turf roots, and a fuel cell.

■ Several institutions (e.g., Tufts, Oberlin, UC Santa Barbara, all Oregon institutions of higher learning) have set a goal of becoming completely climate neutral, typically as part of the American College and University Presidents' Climate Commitment.[8] Oregon State University's plan is to be carbon neutral by 2025. Climate neutrality refers to reducing a campus' net greenhouse gas (GHG) emissions to zero through energy efficiency upgrades and use of alternative energy sources to offset remaining GHG emissions.

■ Coastal Carolina University refurbished a light-industrial warehouse into a science building (an example of *brownfield development*) and installed several energy-saving features, including motion sensors (to shut lights off in unoccupied rooms), $CO_2$ sensors (to match ventilation in classrooms to demand), and energy-recovery wheels (which use exhaust air to preheat or pre-cool incoming fresh air).

■ The University of California at Berkeley, upset that nearly 20 percent of its students were driving cars without additional passengers to campus, initiated a program to encourage carpooling and bicycle use. The result: a 50 percent decrease in single-occupancy cars.

■ Duke University has a comprehensive sustainable dining program, featuring fair-trade and organic food, vegan options, local-grown vegetables, and composting.

**Question 29-2:** By 2012, 658 institutions had signed the Presidents' Climate Committment. Has your institution signed the Presidents' Climate Commitment?[9] If so, what tangible steps has it taken to become climate neutral?

**Question 29-3:** Does your institution have a sustainability office or initiative? If so, visit their website or office and describe a few of the major accomplishments or programs. If not, look online for a nearby institution that has such an office or initiative and answer the question for that institution.

---

[7] More examples can be found at www.aashe.org or www.ulsf.org.

[8] www.presidentsclimatecommitment.org/.

[9] If you do not know, ask an employee of your institution, check its website, or go to www.presidentsclimatecommitment.org/ and follow links to the list of signatory institutions.

# FOCUS: ENERGY USE AT OBERLIN COLLEGE'S ADAM JOSEPH LEWIS CENTER

One of the pioneers in the transformation of college campuses toward sustainability is Oberlin College in Ohio. In 1999, Oberlin broke ground for the Adam Joseph Lewis Center[10], the centerpiece of an ambitious program to reinvent the college as a model of sustainability. At the groundbreaking, David Orr (whom we quoted earlier), professor and director of Oberlin's Environmental Studies Program, stated:[11]

In the century ahead all of those who will be educated here must learn how to:

■ Power society by sunlight and stabilize climate,

■ Disinvent the concept of waste and build prosperity within the limits of natural systems—in ways that can be sustained over the long term,

■ Preserve biological diversity and restore damaged ecosystems, and

■ Do these things while advancing the causes of justice and nonviolence.

We will examine one dimension of this extraordinary building, energy use.

To save energy and reduce greenhouse gas emissions, the Lewis building is equipped with 4,800 ft$^2$ (446 m$^2$) of photovoltaic (PV) panels on the roof.[12] These panels use solar energy to produce 58.65 kW (kilowatts) of electricity. In addition to harvesting solar energy, the Center has energy-efficient lighting, heating, and appliances. The building is also south facing, and constructed with windows that allow light waves to penetrate but insulate against cold air and to passively collect heat from sunlight in the winter. The total cost of installation of the solar array in 2000 was about $386,000, or about $6.60 per watt. This analysis will consider three simple questions: What are the benefits? What are the costs? Is it worth it?

## Calculating the Costs of the Lewis Center

First, let's consider *embodied energy*, which is a cost. Embodied energy refers to the energy that was used throughout the manufacturing process for the solar panels, including extraction of raw materials, industrial processing, transportation, and installation. This embodied energy was from a variety of sources, perhaps including alternative energy, but it is a fair assumption to conclude that most of the energy was from fossil fuels. Thus, it is important to estimate the embodied energy of the solar panels in a cost-benefit analysis.

**Question 29-4:** One published estimate of the energy embodied in PV modules is approximately 5,600 kWh (kilowatt-hour) per 1 kW of output.[13] Calculate the embodied energy for the Lewis Center PV array, which had a rated output of 58.65 kW. Report your answer in MWh (megawatt-hours; 1 MWh = 1,000 kWh).

---

[10] See http://www.nrel.gov/docs/fy06osti/38962.pdf and http://www.coolshadow.com/images/magazines/HPB_Oberlin.pdf.

[11] www.oberlin.edu/ajlc/design_1.html.

[12] This analysis is derived from Murray, M., and J.E. Petersen. 2004. Payback in currencies of energy, carbon dioxide and money for 60 KW Photovoltaic array. Conference Proceedings of the American Solar Energy Society. Used with permission of Dr. Petersen.

[13] Knapp, K., & T. Jester. 2001. Empirical investigation of the energy payback time for photovoltaic modules. *Solar Energy* 71(3): 165–172: Murray, M.E. 2004. Payback and currencies of energy, carbon dioxide and money for a 60 kw photovoltaic array. Senior Honors Thesis Environmental Studies Program Oberlin College, 35 pp.

**Question 29-5:**   A more comprehensive estimate of the embodied energy adds about 40 MWh, which considers the energy to manufacture and install other components (cabling, transformers, inverters, etc.). What is the new embodied energy figure?

**Question 29-6:**   For 2001, the solar arrays produced 51.4 MWh of electricity. Calculate the payback time for electricity using the total embodied energy from Question 29-4 and assuming 51.4 MWh annual electricity production.

Another cost of the solar array system is $CO_2$ production associated with the manufacture of the PV panels. The average carbon dioxide intensity (lbs $CO_2$ per kWh) for the United States, considering the mix of fuels, is 1.45. Table 29-1 shows intensities for different fuels.

To calculate $CO_2$ payback, you need to know $CO_2$ revenues, that is, the amount of $CO_2$ emissions avoided by using solar energy *instead of* electricity from the local utility produced primarily from coal. For Oberlin's supplier, the $CO_2$ intensity was 2.16 lbs $CO_2$ per kWh, resulting in total emissions of 409,800 lbs of $CO_2$.

**Question 29-7:**   What is the $CO_2$ payback time for the Oberlin solar array?

a. First calculate the total annual $CO_2$ revenues in pounds assuming a $CO_2$ intensity of 2.16 lbs $CO_2$ per kWh and total electricity produced of 51.4 MW. (For this calculation, assume that the units for $CO_2$ intensity are kW, not kWh.)

TABLE 29-1   ■   Average carbon dioxide intensities for different fuels in the U.S.

| Carbon dioxide intensity of different fuels (lbs $CO_2$ per k Wh) | |
| --- | --- |
| Coal | 2.117 |
| Oil | 1.915 |
| Natural gas | 1.314 |
| Nuclear | 0.017 |
| Hydro | 0 |

(*Source:* Department of Energy and the Environmental Protection Agency 2000)

b. To determine payback time in years, divide the total emissions (409,800 lbs) by the answer to a.

The final cost we will consider is monetary.

**Question 29-8:** Assume that electricity costs $0.10 per kWh for the lifespan of the panels. What are revenues from the annual 51.4 MWh of solar electricity after twenty-five years? Forty years? After how many years will the initial cost ($385,778) be recovered?

**Question 29-9:** Increase the electricity cost estimate to $0.12 per kWh, the average cost for residential electricity in 2012. Will the initial cost ($385,778) ever be recovered?

**Question 29-10:** So far in this analysis we considered only energy, $CO_2$, and money. Based on our results so far, explain whether you would support placement of solar arrays on a building at your institution. In other words, would it be worth the price Oberlin paid?

GE claims that it can bring the cost of installing photovoltaic systems to under $3 per watt[14] (compared to $6.60 for Oberlin's system).

**Question 29-11:** Except for $CO_2$ emissions, this analysis ignored the environmental and health impacts of energy production (see Issues 7 and 8). Discuss whether your answer to the previous question would have been different had we considered the environmental and health savings associated with solar power production compared to coal or natural gas.

---

[14] http://www.earthtechling.com/2011/11/pv-cost-down-to-3-per-watt-is-ge-goal/.

## Performance of the Lewis Center

One innovation of the Lewis Center is online real-time and historical performance information (Figure 29-2). The graph on the left portion of each figure shows either a monthly (29-2a) or annual (29-2b) curve of solar energy production, energy consumption, and net use, while the right side shows meters registering information for a single day. Net use indicates the difference between consumption and PV production. Negative numbers indicate an excess of solar energy, which is sold back to the utility. The lower right panel in each figure shows the relative location of the Sun.

**Question 29-12:** On June 23, 2006, the Lewis Center solar panels produced 32 kW and the building used 12, whereas on December 18, 2006 they produced 2 kW and the building used 45. Explain the difference in electricity use and production for these two days.

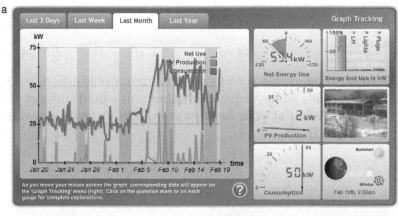

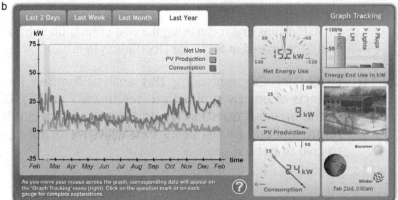

**FIGURE 29-2** Monthly (a) and annual (b) energy performance at the Lewis Center. Real-time and historic performance data are available online at www.oberlin.edu/ajlc/systems_ energy_2.html. (Courtesy of Lucid Design Group.)

## Can We Do Better?

Even though Oberlin College has been a leader in sustainability, there is still more to do. Consider this letter to the editor of the Oberlin student newspaper, the *Oberlin Review*.[15]

Are We Wasting College Money?

To the Editors:

Today as I was walking home . . . I looked in perplexity at the countless number of windows open in every dorm. . . . Arriving at my room just after sunset, I flipped on my light and a brilliant flash dazzled me followed by complete darkness. I asked my RA for a new bulb and was handed a traditional incandescent bulb. What's wrong with this you ask in bewilderment????? First, the college is apparently profusely wasting energy and fossil fuels.

Second, my tuition is obviously paying for it. The college needs take a stance of climate neutrality. . . . Climate neutrality means taking steps that make the college free from the release of greenhouse gases. . . . The effects of climate change have already had far reaching social and economic impacts and these effects will only continue to worsen. As an academic community, it is deplorable that we ignore this. We must use our collective knowledge to expand the teachings of the classroom to the campus and the world. A climate neutral policy would need to include many things. A first and obvious step would be to stop overheating the dorms. This would not only save the earth from the release of greenhouse gases, but also serve to save money. Additionally, for the time being, the college could purchase green power from the grid. Next, the college would have to phase out its use of the huge coal power plant they use to heat the school. Solar arrays on the rooftops of the dorms could compensate for the plant. The college needs to make every building be green, not just the Lewis Center. . . . I urge the Oberlin College and its community to support and implement a climate neutral policy.

**Question 29-13:** Describe your reaction to this letter.

**Question 29-14:** Discuss whether you agree or disagree that college-educated people are responsible for the state of the environment.

**Question 29-15:** Summarize the main points of this Issue.

---

[15] www.oberlin.edu/stupub/ocreview/archives/2002.12.06/perspectives/article18.htm.

# FOR FURTHER THOUGHT

**Question 29-16:** Select one of the institutions that signed the Presidents' Climate Commitment at http://www.presidentsclimatecommitment.org/. Go to its website, and record evidence of tangible progress it has documented toward sustainability.

**Question 29-17:** Become aware of your own energy use. How many outlets are occupied by plugs from your appliances, computers, stereos, rechargers, etc. How many appliances are on all the time? If you don't know, find out what their power use is.

**Question 29-18:** Penn State and others instituted a simple but very effective way to save paper—they changed the margins on word processing programs to 0.75 inches on campus computers.[16] This allowed 19 percent more area for text. Discuss the environmental impact of this practice.

**Question 29-19:** Discuss whether food services on your campus are sustainable. What criteria did you use?

**Question 29-20:** Discuss whether your campus' landscape is sustainable. What criteria did you use?

---

[16] Penn State Indicators Report (2000); Pearce, J.M., and C. F. Uhl. 2003. Getting it done: Effective sustainable policy implementation at the university level. *Planning for Higher Education, 31*(3): 53–61.

# RESTORATION ECOLOGY

## KEY QUESTIONS

- What is an ecosystem?
- What services do ecosystems provide?
- What is the annual value of these services?
- Are there technological replacements for these services?
- Which ecosystems are most endangered?
- What is *restoration ecology*?

## ECOSYSTEM RESTORATION

Humans had negligible impact on the planet thousands of years ago, as hunter-gatherers.[1] Of course there were far fewer of us then. Human impact was minimized because of a nomadic lifestyle that was largely governed by Darwinian natural selection. Early humans hunted where they settled, but, like other top predators in any ecosystem, they selected the less-fit animals and gathered the "low-hanging" (i.e., easy to pick) fruit and vegetables. When it became progressively more difficult to locate and kill animals—because the ones remaining had superior predator-avoidance adaptations—these nomads packed up and moved on. This pattern allowed ecosystems to assimilate the direct human impact (wastes and landscape changes) and let organisms repopulate and to recover relatively quickly. In other words, the ecosystems had the resiliency to naturally restore what minimal damage the human presence caused.

It is an understatement to say that those days are long gone. According to environmental philosopher and provocateur Garrett Hardin, the post-hunter-gatherer human approach became *exploit, ruin, move-on*, which was based on the mistaken assumption that the Earth was too big and its resources too abundant for humans to cause irreversible harm. There are now more than 7 billion of us, and virtually no ecosystem—land or sea—has escaped our footprint. But, we have no place else to "move on."

Fortunately, there is action to undo some of the damage humans have wrought on the planet. This is the movement known as *ecological restoration*. The Society for Ecological Restoration defines ecological restoration as "the process of assisting the recovery of an ecosystem that has been degraded, damaged, or destroyed."[2]

Ecological restoration can be conducted on a small scale (like a backyard or a park) or a large scale (e.g., Great Plains, Everglades, San Francisco Bay Delta, Great Lakes) (Figure 30-1). The idea of ecological restoration is also built into the Endangered Species

---

[1] Pimentel, D., & M. Pimentel. 1996. *Food, energy, and society*. (University Press of Colorado), p. 363.
[2] Society for Ecological Restoration, www.ser.org .

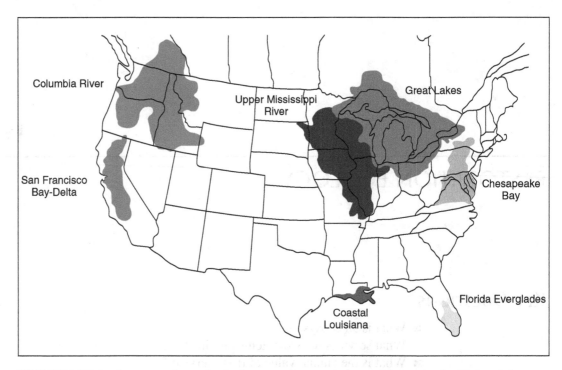

**FIGURE 30-1**  Large-scale ecological restoration projects underway in the Unitesd States. (Courtesy of North Mid-West Institute.)

Act of 1973, which requires that "the ecosystems upon which [threatened and endangered species] depend" be protected.

## Ecosystems and Ecosystem Services

The United Nations Convention on Biodiversity defines an ecosystem as "a dynamic complex of plant, animal, and microorganism communities and their nonliving environment interacting as a functional unit."[3] Ecosystems can be as large as the entire planetary biosphere or as small as a drop of water. Ecosystems are usually classified by the dominant species or type of physical environment (e.g., pine forest ecosystem or deep-sea hydrothermal vent ecosystem).

From a human perspective, ecosystems provide numerous services (Table 30-1) for which technological substitutes are either lacking, excessively expensive, or undesirable. According to the Ecological Society of America (ESA), *ecosystem services* are "the processes by which the environment produces resources that we often take for granted such as clean water, timber, and habitat for fisheries, and pollination of native and agricultural plants."[4]

In a paper published in *Nature* in 1997, environmental economist Robert Constanza and colleagues estimated the annual dollar value of these ecosystem services at between $16 and $54 *trillion* globally.[5] They found coastal environments to have disproportionately high value, contributing 43 percent of the value of ecosystem services while occupying only 6.3 percent of the Earth's surface.

---

[3] http://www.cbd.int/ (Convention On Biological Diversity), as cited in K. E. Vigmostad, N. Mays, A. Hance, and A. Cangelosi. 2005. Large-scale ecosystem restoration: Lessons for existing and emerging initiatives, Northeast Midwest Institute.

[4] Ecological Society of America, www.actionbioscience.org/environment/esa.html.

[5] Costanza, R., R. d'Arge, R. de Groot, S. Farber, M. Grasso, B. Hannon, K. Limburg, S. Naeem, R. V. O'Neill, J. Paruelo, R. G. Raskin, P. Sutton, & M. van den Belt. 1997. The value of the world's ecosystem services and natural capital. *Nature 387*. pp. 253–260.

**TABLE 30-1** ■ Estimate of the value of ecosystem services from Costanza *et al.* 1997.

| Ecosystem Service | Value (trillion $US) |
|---|---|
| Soil formation | 17.1 |
| Recreation | 3.0 |
| Nutrient cycling | 2.3 |
| Water regulation and supply | 2.3 |
| Climate regulation (temperature and precipitation) | 1.8 |
| Habitat | 1.4 |
| Flood and storm protection | 1.1 |
| Food and raw materials production | 0.8 |
| Genetic resources | 0.8 |
| Atmospheric gas balance | 0.7 |
| Pollination | 0.4 |
| All other services | 1.6 |
| **Total value of ecosystem services** | **33.3** |

**Question 30-1:**  Express 54 trillion as a number (i.e., using 0s and commas)

**Question 30-2:**  At the date of publication of the Nature article (1997), U.S. gross national product (GNP), a measure of the values of goods and services in the United States, was $8 trillion.[6] How did the value of ecosystem services compare to GNP?

**Question 30-3:**  In their Nature article, Costanza et al. write, "to replace the services of ecosystems, one would need to increase global GNP by at least $33 trillion." Discuss whether it makes more sense to devote resources to ecosystem sustainability and restoration or to spend funds to develop new technologies to replace ecosystem services. What assumption(s) about the Costanza article did you use to answer this question?

---

[6] U.S. Bureau of the Census, www.census.gov/compendia/statab/.

Ecosystems are endangered all over the planet, largely due to habitat loss, degradation, and fragmentation. A partial list of endangered ecosystems would include forests (boreal, temperate, tropical), wetlands (fresh and salt water), coral reefs, estuaries, shrublands, and grasslands—indeed, it would be difficult to name a type of ecosystem not under severe stress. In light of the declining state of ecosystems and their economic and ecological value, ecological restoration has become a necessity.

We will examine one of the nation's most endangered (and overlooked) ecosystems, the longleaf pine system.

## The Longleaf Pine Ecosystem

Longleaf pines (*Pinus palustris,* Figure 30-2) are one of the four main types of softwoods (shortleaf, loblolly, and slash pines are the others) in the southeastern United States. They grow to 30 to 35 meters and typically live 100 to 150 years (and up to 500 years). Longleafs have been called the South's "hardest working" softwood and are used intensely by the construction and naval stores industries.[7]

Before European settlement, longleafs prospered throughout the South, at one time covering 94 million acres (Figure 30-3).[8] Longleaf pine ecosystems are *open* and *park-like* and are characterized by large numbers of pine needles on the forest floor. These needles serve as a fuel source for the fires that are a necessary part of the longleaf's life cycle. Fire serves numerous functions in longleaf ecosystems. These include aiding the growth of seedlings, excluding invasive plants, releasing nutrients, promoting seed production by native grasses, reducing competition from hardwoods, and reducing understory vegetation. Excessive understory vegetation can serve as fuel for fires that can burn out of control, destroying the sensitive overstory of a forest.

Longleaf ecosystems provide habitat for numerous endangered and otherwise rare species, including the endangered red-cockaded woodpecker, gopher tortoise, and fox squirrel, as well as game animals like white-tailed deer, wild turkey, and northern bobwhite.[9] These ecosystems also support an extremely rich diversity of understory plants.[10]

**FIGURE 30-2** Adult longleaf pines. Longleafs can be distinguished from other pines by these features: thick, reddish-brown, scaly bark; long, dark-green needles in bundles of three; small, irregularly shaped crown. (D.Abel)

---

[7] *Naval stores* refers to nonwood products (e.g., turpentine) obtained from pine trees.
[8] Brockway, D.G., K. W. Outcalt, D. J. Tomczak, & E. E. Johnson. 2005. Restoration of Longleaf Pine Ecosystems. USDA Forest Service General Technical Report SRS-83.
[9] Ibid.
[10] Ibid.

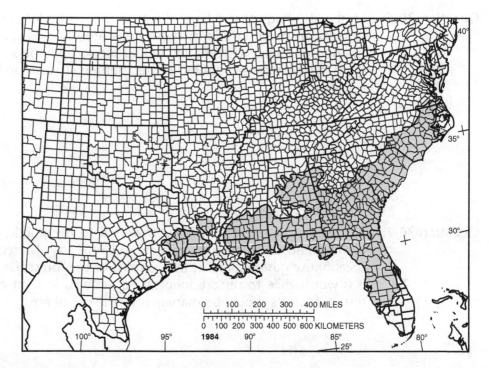

**FIGURE 30-3** Native range of the longleaf pine in the southeastern United States. (Source: www.sfrc.ufl.edu/4h/Longleaf_pine/pinpal1s.jpg)

By 1995 longleaf pine habitat had been reduced to about 2.9 million acres. Most of the loss occurred between 1750 and 1850, with virtually all old-growth longleaf removed by 1930. Primary causes of the destruction include logging, conversion to agriculture, conversion to pine plantations (primarily loblolly, a native of west Texas), grazing by feral hogs on young seedlings, and fire suppression.

**Question 30-4:** What percentage of the original area of longleaf pine ecosystems remained in 1995?

**Question 30-5:** In Florida, from 1987 to 1995, 3,000 hectares per year of longleaf pine were converted to urban uses and 1,500 hectares per year to agriculture. Discuss what kind of action, government or other, you would support to reduce loss or promote restoration of longleaf ecosystems. Who should pay for such restoration?

**Question 30-6:**   Feral (wild) hogs pose a large threat to the reestablishment of longleaf pine ecosystems. In many locations, these hogs are diseased and unfit for human consumption. Discuss whether you would support a large-scale hog eradication program (typically carried out by licensed hunters) to assist in longleaf pine ecosystem restoration.

**Question 30-7:**   The subject of forest fires is controversial to the public, who see forest fires as harmful and destructive to wildlife and property. Moreover, forest fires release greenhouse gases stored in the biomass of the forest. Discuss whether you think it is worthwhile to restore longleaf ecosystems in light of the requirement that restored forests should be managed by the use of fires.

**Question 30-8:**   Hexazinone herbicide is used to accelerate the restoration of longleaf pine systems. Pine trees are tolerant, and the herbicide seems to be relatively safe for terrestrial organisms, but it may be toxic to aquatic organisms. Discuss whether the benefit of accelerating the restoration of longleaf pines outweighs the use of this herbicide. What additional information would you like to have to answer this question?

Financially, it is more costly to restore a longleaf forest than to plant loblolly or other softwoods. However, according to Brockway et al., "Once established, longleaf pine becomes a low risk investment that is easily managed with prescribed fire, resistant to most pathogens and insects, and inexpensively renewed through natural regeneration. The superior growth form and wood quality of longleaf pine also produces excellent sawtimber."[11]

**Question 30-9:**   Discuss whether the purpose of restoration should be economic or ecologic or both. If both, which is the more important purpose?

---

[11]  Ibid.

**Question 30-10:** Would you support a government program to restore public and private longleaf forests that eventually could be clearcut for forestry products? Explain.

**Question 30-11:** Summarize the main points of this issue.

## FOR FURTHER THOUGHT

**Question 30-12:** After a wildfire destroyed 52,000 acres of pine and fir trees in southeastern Washington, the U.S. Forest Service reacted in a way that has been called "a quiet revolution in federal wildfire restoration"[12]—they scattered 21,000 pounds of native grass seed. U.S. Forest Service spokeswoman Joani Bosworth said "this planting . . . likely exceeds all the post-fire native seedlings combined nationally."[13] The native grasses, which include Idaho fescue, bluebunch wheat grass, prairie junegrass, and blue wild rice, are fire- and weed-resistant. While nonnative seeds cost less to buy, they resulted in more wildfires and insect infestations because they had not evolved where they were planted and thus were not well-adapted to the climatic and geological conditions (like the soil).[14] Would you support regulations requiring all restored areas to use native plants? Why or why not?

**Question 30-13:** Half a billion people live within 100 kilometers (62 mi) of coral reefs. The report Status of Coral Reefs of the World: 2004 by the Australian Global Coral Reef Monitoring Network documents the impacts humans have on the global coral reef crisis.[15] In addition to natural forces with which corals have evolved for millions of years, the report lists these major anthropogenic (human-caused) stresses to coral reefs:

- Sediment and nutrient pollution from the land
- Overexploitation and damaging fishing practices
- Engineering modification of shorelines
- The global threats of climate change causing coral bleaching, rising sea levels, and acidification, threatening the ability of corals to form skeletons.

One of the greatest threats to corals comes from increasing water temperature associated with global climate destabilization. This is because, like virtually all living things, corals live closer to the highest temperature they can tolerate, rather than the lowest. Corals

---

[12] Cockle, R. Ravaged by fire, land returns to its roots. *The Sunday Oregonian*, June 16, 2006, p. B1.
[13] Ibid.
[14] Ibid.
[15] Wilkinson, C. (ed.) 2004. *Status of Coral Reefs of the World: 2004.* Australian Institute of Marine Science.

cannot readily migrate, nor can they adapt to such a rapidly changing ocean temperature. Although challenging, restoration of coral reef ecosystems must include addressing sediment and nutrient pollution from the land, over-exploitation and damaging fishing practices, and engineering modification of shorelines as well as destructive land-use practices such as deforestation.

Discuss whether you think it is worthwhile to attempt to restore coral reef ecosystems until the issue of global climate destabilization is addressed seriously at the national or international level. Who is responsible, if anyone, for ensuring the survival of coral reefs or paying for the restoration? Explain your answer.

**Question 30-14:** Access the web site of the International Coral Reef Initiative (ICRI:http://www. icriforum.org/) and report on its mission to restore coral reefs.

# INDEX

Accuracy, reasoning, 9
Acid precipitation, 126
Addition, using scientific notation, 5–6
Africa
    AIDS in, 35
    bush meat crisis, 199
    Mt. Kilimanjaro, 19–20
Agriculture
    climate change, 63
    grains and origin of, 176–177
    land-use, 133
    oil, 91
    practices reducing soil loss, 188–190
    soil concerns, 192–193
    sustainable, 18
    water pollution in China, 114
Agro-ecosystems, 177
AIDS, impact on population growth, 35
Air pollution
    Chesapeake Bay, 209, 212–213
    coal, 81
    coal combustion, 84, 85
    leaf blowers, 140
Air Resources Board (ARB), 136
Aliphatics, 169
Allergies, 136, 138
Alloying, 157
Aluminum
    cans, 273–275, 276
    electricity and, cans, 276
    mining and refining, 273
    production, 275
    recycling, 273–276
    refining and electricity costs, 275–276
    subsidies, 275
Amalgamation, gold mining, 162–163
American Farmland Trust (AFT), 185
American Geological Institute, 191
American Wind Energy Association, 99
Anaerobic microorganisms, 55
Anthracite, 82
Anti-noise legislation, 141–142
Aquaculture, 18, 235–237

Arch Coal, 88
Arctic National Wildlife Refuge (ANWR),
    oil reserves, 75
Aromatics, 169
Association for the Advancement of
    Sustainability in Higher Education,
    289
Assumptions, 11
Atmosphere, 53–54
    carbon dioxide in, 57–59
    changes caused by humans, 56–57
    current composition, 56
    development of current, 55
    early planetary, 55
    functions of, 55
    greenhouse effect, 56–57
    weighing, 54–55
Atmospheric pressure, 54
Australia, water use, 113

Ballast water, 218–220
Bangladesh, 7, 43–45, 47
    flooding in, 43–45
    population growth, 45–46
Basel Accords on Toxic Waste, 28
Basel Action Network, 270
Basel Convention, 27, 28
Bauxite, 273
Bed Bath & Beyond, 265
Better Site Design, 242
Bioaccumulate, 159
Bioaccumulation, 171–172
Biodiversity, 21
Biomagnification, 170, 171–172, 173
Biomass, 202
Biotic potential, 37
Bitumen sands, 80
Bituminous coal, 82
Bonneville Power Authority (BPA), 96,
    275–276
Boreal Forest, 80
Boreal Forest Network, 204
Boreal forests, 203–205

Breadth, of reasoning, 10
Breast feeding, persistent organic
    pollutants and, 174
Bronze, 157
Brownfield development, 290
Buildings. *See also* High performance
    building
    high-performance, 255–256
    sustainable, 254–264
Bush meat crisis, Africa, 199
Business reply envelope (BRE), 207
Bycatch, 230, 231
Bykill, 230, 231

California
    climate change, 112, 113
    water demand, 112, 113
    water use in, 111–112
California Energy Commission, 257,
    259
Campus Ecology Program, 289
Canadian boreal forests, 204–205
Canadian tar sands, 80
Carbonate, 55
Carbon dioxide, 56, 122
    atmosphere, 57–59
    emissions factors by state, 262
    intensities for different fuels, 292
    soil and, 191–192
Carbon monoxide (CO), 126
Carbon sequestration, 88
Carcinogenic, 170
Carcinogens, 126
Carrying capacity, 37–38
    assessing, 39
    impact of population, 38
Cathode-ray tubes (CRTs), 270
Cell phones, 152, 270
Census Bureau, 6
Center for a New American Dream, 259
Center for Disease Control and Prevention
    (CDCP), 23, 88
Cesspools, 285

Chesapeake Bay
air pollution and, 212–213
airshed, 209
background of, 208–209
bay remediation, 214–215
Commission (CBC), 211, 282
environmental research, 210–212
estuary restoration, 208–216
invasive species and, 220–227
*Pfiesteria* and manure, 213–214
sewage impact on, 212
watershed, 209
Chesapeake Bay Ecological Foundation
(CBEF), 211
Chesapeake Bay Foundation (CBF), 211,
214, 215
China
automotive use, 126–129, 130
externalities of coal use in, 82
food availability, 176
global air pollution, 89
population growth, 32
Three Gorges Dam, 95
U.S.-China trade deficit, 81
water, 108, 113–114
Chlorination, 169–174
highly exposed populations, 172
organochlorines, 168–169, 170
polychlorinated biphenyls (PCBs),
170–172
Chlorofluorocarbons (CFCs), 22, 53
Chromated copper arsenate (CCA), 145
*Citizens United v. Federal Election
Commission* (2010), 16
*City of Philadelphia v. New Jersey* (1978),
150–151
Clarity, of reasoning, 9
Climate, 55
Climate change
California, 112, 113
greenhouse gases and, 60–68
Gulf Stream, 64–65
impacts of global, 62–64
North Atlantic circulation, 64
sustainable, 21–22
Climate neutral, 202
Coal, 81–89
air pollution and, 81
bed methane, 84
clean, 89
electricity generation, 98
externalities of use, 82
global recoverable reserves, 87
global use of, 86–88
origin and nature of, 82–83
pollution from, 84
as resource, 86
Coalition to Reform Animal Factories,
280, 285
Coastal Carolina University Campus and
Community Sustainability Initiative,
289

Coastal lands
climate change, 63
sustainable development, 247–253
Coastal population growth, 247–248
designs, 249, 250
environmental indicators of, 251
industrial hog production, 282–284
model, 248–252
Coefficient of thermal expansion (CTE), 66
Commercial forests, 20
Communities, sustainable, 239–246
Compassion Over Killing, 281, 286
Competitive Enterprise Institute, 276
Compound growth equation, 6–7
Confined animal feeding operations
(CAFO), 29, 280–281, 285
Congressional Research Service (CRS),
147, 148
Conservation design, coastal growth,
249–251
Conservation Reserve Program, 185
Construction and demolition waste (C&D),
22, 144–148
categories, 145
municipal solid waste (MSW), 145–148
pressure-treated wood and, 145
Consumption
fish, 84, 85, 229–230
sustainable, 22–23
Consumptive use, 110, 111
Contaminated water, 109–110
Conventional design, coastal growth,
249–251
Conveyor-belt circulation, 65
Coral reefs, 303–304
Corporate Average Fuel Economy (CAFE)
standards, 123–124
Corporations, assumptions about, 15
Critical thinking, 1, 8–16
accuracy in, 9
breadth, 10
clarity in, 9
defined, 8–9
depth, 10
first order thinking, 8
global grain production, 177–179
intellectual standards in, 9–11
logical fallacies, 11–12
logic in, 19
precision, 9
relevance, 9–10
second order thinking, 8
Crude birth rate (CBR), 48
Crude oil, 70
*Cryptosporidium*, 108
Cyanide, 159
Cyanide leaching, gold mining, 159–162
Cyanobacteria, 55
*Cynodon*, 138

Day of 6 Billion, United Nations, 31
Death rate, in coal mining, 82

Deepwater Horizon, 96
Deforestation, 23, 44, 198–199
greenhouse effect and, 53, 56
Deltas, 45
Density, population, 34, 45, 240
Department of Pesticide Regulation, 14
Depth, of reasoning, 10
Desertification, 186–187, 190
Development
defined, 23–24
environmental impact of trade and, 27
sustainable, 23–24
Diet, sustainable, 277–286
meat consumption, 279–282
myplate.gov, 277–279
Dioxins, 269
Direct Marketing Association, 205
Dirty Dozen, 167
Disposable paper cups, impact on forests,
205–206
Distributaries, 45
Division, using scientific notation, 5
Dogwood Alliance, 202
Doubling time, 7
energy consumption, 97
formula, 32–33
population growth, 7–8, 208, 240, 241
Downsizing, 123
Dust Bowl, 185, 186

EarthCraft House, 256
Earth Policy Institute, 177, 178, 236
Ecological economics, 23
Ecological footprint, 40–41
Ecological Society of America (ESA), 298
Ecology, restoration, 297–304
Economics, wind energy, 101
Ecosystem
longleaf pine, 300–303
restoration, 297–298
services, 298–303
Education. *See* Higher education
Eggs, 286
Electricity
aluminum cans, 276
aluminum refining, 275–276
nonhydroelectric renewable, 97
sources, 98
sustainable sources, 98–99
Electronic waste, 152–154
Embodied energy, 291, 292
Emissions, 122–123, 125–126
carbon dioxide, 262
leaf blowers, 139–140
motor vehicles, 122–123, 125–126
Endangered Species Act (ESA), 12,
297–298
Endocrine disrupters, 167
Energy
calculating costs of Lewis Center,
291–293
consumption, 96–99

embodied, 291, 292
renewable, 95, 96, 97, 101–103
sustainable, 21, 95–103
wind, 99–101
Energy Information Administration (EIA), 91, 100, 260
Energy Policy and Conservation Act of 1975, 123
Enrichment, 156
Environment
  assumptions about, 12
  CAFÉ standards and, 123–124
  impact of development on, 24
  international agreements, 28
  Kuznets curves, 24–28
  math, use of, 6–8
  motor vehicles and, 118–124, 125–131
  oil refining, impacts on, 75–76
  trade and development, 27
Environmental Action Coalition, 289
Environmental Defense, 68, 174, 232, 253
Environmental Kuznets curves (EKC), 25, 26
Environmental Protection Agency (EPA), 14, 210
Environmental quality, 23
  global trade and, 26
Environmental research, Chesapeake Bay, 210–212
Environmental resistance, 37
Environmental subsidies, 192
Environmental Sustainability Index (ESI), 18
Equity, sustainable, 22
E-recycling, 153, 154
Erosion, soil, 185–186
Estuaries, restoring, 208–216
Ethical Gourmet, 278–279
Eureka Recycling, 271
European Union (EU), 18, 93
Evaporation, 106, 107
Evidence, 10
Exotic species, 217, 218
Exponential growth, 31–34
Externalities, 15, 192, 275
  coal use, 81, 82
Exxon Valdez, 96

Federal Noise Control Act of 1972, 141
Federal Trade Commission (FTC), 14
Fertile Crescent, 176, 187
Fertility, 48, 49
First order thinking, 8
Fish consumption, mercury advisories, 84, 85
Fisheries/fishing
  aquaculture, 235–237
  bycatch, 230, 231
  bykill, 230, 231
  catch analysis, 233–237
  environmental costs of, 230–231

European, 228–229
global, 228–238
importance of, 229
Patagonian toothfish, 231–233
slimeheads, 231–233
sustainable, 18–19, 229–230
Flooding, in Bangladesh, 43–45
*Flow* (2008), 117
Flue gas desulfurization (FGD), 88
Food and Agriculture Organization (FAO), 229
Food and Drug Administration (FDA), 14
ForestEthics, 207
Forests. *See also* Global forests
  boreal, 203–205
  climate change, 63
  southern, 201–203
  sustainable, 19–20
  temperate, 200–201
  tropical, 197–200
Forest Stewardship Council (FSC), 207
Fossil fuels, 57, 280
  greenhouse effect and, 57
  oil and natural gas, 69–80
Fracking, 78, 101, 103
Free market principles, 109
Fresh Kills landfill, 149–150

Garbage, Supreme Court rulings, 150–151.
  *See also* Waste
Gasoline
  consumption, 128–129
  emissions, 125
  global reserves, 77
  prices, 120, 128
Gas powered leaf blowers, 138–140
Generalization, hasty, 11
Global economic product (GEP), 90
Global forests
  boreal, 203–205
  classes of, 197–205
  disposable paper cups and, 205–206
  junk mail and, 205
  protection of, 196–197
  southern, 201–203
  temperate, 200–201
  tropical, 197–200
Global population growth, 29–36
Global trade, environmental quality and, 26
Global warming, 45
  sea-level changes and, 66–67
Global water use, 105–106
Gold
  prices, 158
  production, 157–159
  recycling, 163–165
Gold mining, 156–165
  amalgamation method, 162–163
  cyanide leaching, 159–162
  modern methods of, 159–163

Gold potassium cyanide (GPC), 158
Golf courses, 143
Government, role of, 12–16
Grain consumption, 176–177
  in China, 176
  in United States, 176
Grain production, 175–183
  critical thinking, 177–179
  global prices, 180
  grainland area, 179–180
  topsoil loss, 189
  world, 178
Green building. *See also* High performance building
  economics of, 256–259
  residential, 260–263
Green consumerism, 24
Greenhouse effect, 21, 56–57
Greenhouse gases, 56, 290
  atmosphere, 53–59
  climate change and, 60–68
Green movement, 24
Greenpeace, 174
Green Power Network, 103
Green revolution, 114, 177, 179
Gross domestic product (GDP), 25
Groundwater, 106, 107
  West Bank, 115
*GSA Today*, 64, 65
Gulf of Mexico, 94, 280
Gulf Stream, 64–65

Haiti, 50, 108
Hawaii, renewable energy, 101–103
Health
  climate change impacts, 63
  nutrition and, 23
Heat index, 62
Heating, ventilation and cooling (HVAC), 260
Higher education
  defined, 287–290
  energy use at Oberlin College, 291–295
  sustainability, 288–289
  Talloires Declaration, 288–289, 290, 296
High performance building, 255–256
  EarthCraft certification, 256
  economics of green building, 256–259
  LEED certification, 256
Highway safety, trash hauling and, 151
Hog farming
  coastal population growth, 282–284
  sustainability of, 284, 285
Home building, 243–245
Home Depot, 265
Housing
  residential green building, 260–263
  sustainable, 254–264
Hubbert's Curve, 79
Human carrying capacity, 38

Humane Methods of Slaughter Act, 281
Human health, 62
Hurricane Agnes, 243
Hurricane Katrina, 253
Hurricanes, 42, 242–243
    climate change, 61
    soil erosion, 185
Hybrid motor vehicles, 118
Hydraulic fracturing, 78, 101
Hydrologic cycle, 106, 107

Icehouse, 21
Illegal immigrants, 217–218
Immigration, illegal, 217–227
    illegal immigrants, 217–218
    invasive species along Pacific Coast,
        220
    invasive species and Chesapeake Bay,
        220–227
Impervious surface, 242
    coastal development, 251
    home building and, 243–245
Incinerators, 144, 147, 267, 269, 283
India
    coal use, 81
    population growth, 32
Indonesia, 129, 130, 196, 230
Industrial Revolution, 22, 56, 57
Industry, sustainable, 22
Infiltration, 106, 107
Instant lawn, 135
Intellectual standards
    critical thinking framework, 10–11
    solid reasoning, 9–10
Intensification, 177
Intergovernmental Panel on Climate
        Change (IPCC), 56, 62
    3rd assessment, 60–61
    4th assessment and beyond, 61–62
International agreements
    Basel Accords on Toxic Waste, 28
    environmental protection, 28
    Montreal Protocol, 22, 28
International Food Policy Research
        Institute (IFPRI),
        190, 191
International Monetary Fund (IMF), 90
Invasive species, 26
    Chesapeake Bay, 220–227
    Pacific Coast, 220
    veined rapa whelk, 221–224
    zebra mussel, 224–226
Israeli/Palestinian water conflict, 114–116

Jeffersonian principles, 12–13
Junk mail, 205, 207
Justice and equity, sustainable, 22

Keeling Curve, 57
Keystone XL Pipeline, 80
Kuznets curves, 24–28

criticisms, 26
shapes of, 25

Lagoons, 282, 285
Landfills, transporting trash, 148–150
Land-use, North America, 133–134
Large-scale mining, 24
Laterites, 185
Lawn care industry, 137
Lawrence Berkeley National Laboratory,
        257
Leaf blowers
    emissions from, 139–140
    gas powered, 138–140
    noise, 141
    resuspended dust, 140
LEED (Leadership in Energy and
        Environmental Design), 256, 259,
        290
Lignite, 82
Lipophilic, 171
Logic, of reasoning, 10
Logical fallacies
    ad hominem, 12
    appealing to pity, 12
    appealing to tradition, 12
    appeal to deference, 12
    appeal to popularity, 12
    composition, 11
    confusing coincidence, 12
    critical thinking, 11–12
    false choice, 12
    hasty generalization, 11
    irrelevant conclusion, 12
    rigid rule, 12
    starting with answer, 11
Longages, 114
Longleaf pine, ecosystem, 300–303
Low-impact development, 253

Manufacturing, sustainable, 22
Mariculture, 235–236
Marine Stewardship Council, (MSC), 278
Math
    critical thinking and, 1–16
    doubling time, 7–8
    projecting population growth and, 6–7
    use in environmental issues, 6–8
McDonald's, paperboard, 206
Meat
    consumption, 279–282
    production, 181–183, 279, 280
Megawatts (MW), 99
Mercury
    fish advisories, 84, 85
    gold mining, 159, 162–163
    synergism with PCBs, 172
Meriodonal overturning circulation, 64
Methane, 56, 57, 84, 280
Methylmercury, 126, 159
    biomagnification, 173

contamination, 84, 85
Metric system, 1–3
    conversions, 2, 3
    prefixes and equivalents, 2
    units, 2
Migration, population growth and, 48–52
Mineral deposit, 156–157
Mining. *See also* Aluminum; Gold mining
    large-scale, 24
Monoculture, 134
Monocultures, 20, 201
Monopolies, 109
Montreal Protocol, 22, 28, 187
Motor vehicles
    CAFE standards, 123–124
    emissions, 122–123, 125–126
    environment and, 118–124
    fuel economy, 123–124
    global trends, 130–131
    sales of, 118–119, 120, 121
    social cost of, 191–200
    use in China, 126–129
Mountaintop removal, 89
Mt. Kilimanjaro, 19–20
Multiplication, using scientific notation, 5
Municipal solid waste (MSW), 22, 25,
        145–148
    categories of, 145
    exporters and importers, 148
    recycling, 147, 267
    transporting trash, 148–150
Mutagenic, 170
*Mycroscystis*, 211
MyPlate.gov, 277–279

Natality, 48
National Academy of Sciences, 136, 230
National Aeronautics and Space
        Administration (NASA),
        19, 80, 198
National Association of Home Builders
        (NAHB), 256, 260
National Invasive Species Council (NISC),
        217
National Mining Association, 159
National Oceanic and Atmospheric
        Administration
        (NOAA), 68, 219, 238, 248
National Renewable Energy Laboratory,
        255
National Research Council (NRC), 51, 52
National Safety Council, 122, 270
National Wildlife Federation Campus
        Ecology Program, 289
Natural gas, 70
    deposits of, 76
    electricity generation, 98
    global gas reserves, 77
    hydraulic fracturing, 78
    Prudhoe Bay, 78–79
    transporting, 77

Natural habitat, climate change, 63
Natural increase, 49
Natural Resources Defense Council, 119
New Urbanist design, coastal growth, 249–251
New York City, recycling, 149
Nitrogen oxides (NOx), 24, 85, 122, 126
Noise pollution, 141–142
Noisette Company, 253, 254
Noisette Project, 254–255, 257
North America, land-use changes, 133–134
North Atlantic Circulation, climate change and, 64
Nuclear energy, electricity generation, 98
Nutrient enrichment, 210
Nutrient mining, 191
Nutrition, health and, 23

Oberlin College, 287
    calculating cost of Lewis Center, 291–293
    embodied energy, 291, 292
    energy use, 291–295
    performance of Lewis Center, 294
    sustainability, 295
Oil, 69–70
    agriculture, 91
    ANWR's oil reserves, 75
    consumption, 91–93
    crude, 70
    distribution of, 70
    extraction of, 70
    fields, 72
    origin of, 70
    peak, 69
    price, 71
    producing and transporting petroleum, 94
    proven oil reserves, 70–72
    refining, 75–76
    reserves remaining, 70–72
    sands, 80
    world oil demand, 73
Orcas, 167, 171
    polychlorinated biphenyls and, 172–173
Ores, 156
Organic Trade Association, 183
Organization for Economic Co-operation and Development (OECD), 27
Organochlorines, 168–169
    aliphatics, 169
    aromatics, 169
    safety of, 170
*Our Changing Planet*, Public Television, 41
Oxidation of pyrite, water pollution and, 84
Ozone, 126
Ozone-depleting chemicals (ODCs), 28
Ozone layer, 55

Pacific Coast, invasive species along, 220
Paperboard, 206
Patagonian toothfish, 231–233
Peak oil, 79
Peat, 82
Pedalfers, 185
Pedocals, 185
People's Republic of China (PRC). *See also* China
    trade deficit, 81
    water and development in, 113–114
Permafrost, 76, 107, 203
Persian Gulf, 94
Persistent Organic Pollutants (POPs), 22, 166–174
    agricultural soils, 168
    convention on, 166
    costs of exposure to, 168
    exposed populations, 172
    global concern, 167, 168
    organochlorines, 168–169
    precautionary principle and, 168–169
Petroleum, 70. *See also* Gasoline; Oil
    transporting, 94
*Pfiesteria*, 213–214
*Pfiesteria piscicida*, 136
*Pinus palustris* (longleaf pine), ecosystem, 300–303
Planetary atmospheres, 55
Plastic, future of, 271–272
Plating salts, 158
Point of view, 10
Poison runoff, 119
Pollution, 21
    air, 81, 84, 85, 140
    noise, 141–142
Polychlorinated biphenyls (PCBs), 170–171
    bioaccumulation, 171–172
    biomagnification, 171–172
    and orcas in Puget Sound, 172–173
    synergism with mercury, 172
Population
    carrying capacity, 38
    density, 45
    sustainable, 23
Population growth, 239–240
    AIDS impact, 35
    Bangladesh, 45–46
    coastal, 42–47, 247–248, 282–284
    coastal growth model, 248–252
    compound growth equation, 6–7
    density, 34–35
    doubling time, 7–8, 32–33, 240, 241
    exponential growth and impacts, 31–34
    global, 29–36
    hogs, 282–284
    migration and, 48–52
    Prince William County, 240–241
    sample growth calculation, 6–7
Post-consumer recycling, 268–269

Post-consumer waste (PCW), 269
Prairie potholes, 63
Precautionary principle, 12, 14–15, 168–169
Precipitation, 106, 107
Precision, of reasoning, 9
Pre-consumer recycling, 268–269
Pressure, atmospheric, 54
Pressure-treated wood, 145
Prince William County, Virginia
    growth in, 240–241
    home building in, 243–245
    housing data impervious surface area, 243
    hurricanes and tropical storms in, 242–243
Professional Lawncare Network (PLN), 135, 136
Progressive Policy Institute, 27
Projecting, 240
Proven oil reserves, 70–72
Prudhoe Bay, natural gas from, 78–79
Public subsidies, 192
Puget Sound, polychlorinated biphenyls in, 172–173
Purpose, 10–11
Pyrite oxidation, 84

Quality, water, 108
Quality of life, 91, 94

*Rapana venosa* (veined rapa whelk), 221–224
Rebuilding Together New Orleans, 253
Recessive gene, 177
RecycleBank, 265
Recycling, 265–268
    aluminum, 273, 273–276
    e-waste, 270–271
    gold, 163–165
    New York City, 149
    plastics, 271–272
    pre-consumer vs. post-consumer, 268–269
    reduction and reuse, 271
Reduction, 271
Relevance, of reasoning, 9–10
Remediation, Chesapeake Bay, 214–215
Renewable energy
    electricity generation, 98
    Hawaii, 101–103
    technologies, 95
    wind, 95, 99–101
Responsible Purchasing Network (RPN), 259
Restoration ecology, 297–304
Resuspended dust, 140
Reuse, 152, 271
Rigid rule, fallacy of, 12
Risk Transition concept, 113
Rocky Mountain Institute, 257

Romania, cyanide leaching, 160–162
Royal Commission on Environmental
    Pollution, 228
Runoff, 106, 107
Russia, boreal forests in, 203–204

Salinization, 186
Sanford Principles, 254, 255
Santa Clara County v. Southern Pacific!
    (1886), 15
Scientific notation, 4–6
    addition using, 5–6
    division using, 5
    multiplication using, 5
Sea-level
    areas affected by, 67
    assessment, 61–62
    Bangladesh, 43, 45
    global warming and, 66–67
    impact of climate change, 63
Seawater, thermal expansion of, 66–67
Second order thinking, 8
Sewage
    Chesapeake Bay, 212
    lagoons, 282–283
Shortages, 114
Sierra Club, 68, 88, 188, 246, 276, 289
Slimeheads, 231–233
Smithfield Packing Company, 284
Society for Ecological Restoration, 297
Soil
    accelerating erosion, 185–186
    carbon dioxide and, 191–192
    concerns over, 192–193
    degradation of, 190–191
    desertification and salt buildup,
        186–187
    nature of, 184–185
    practices reducing loss, 188–190
    toxins in, 187–188
    urbanization and soil loss, 185
Solar energy, 96, 101, 102, 294
Soot, 122
Source bed, 70
Source reduction, 271
Southern forests, 201–203
Southface Energy Institute, 260
Species, climate change, 63
Sprawl development, 136, 185, 239
Stainless steel, 157
Standard of living
    estimates, 90–91
    using oil consumption to estimate,
        91–93
    world, 90–93
Starbucks
    disposable paper cups and, 205–206
    recycling, 265, 266
Steel, 157
Storm surges, 45
Stormwater management, 242

Sub-bituminous coal, 82
Submerged aquatic vegetation (SAV), 210
Subsidence, 45
Subsidies, 156, 192, 275
Suburbanization, 133
Sulfur dioxide, 65
Supreme Court, rulings on garbage,
    150–151
Surface mining, 89
Sustainability
    of agriculture, 18
    biodiversity and, 21
    buildings and housing, 254–264
    climate change and, 21–22
    coastal development, 247–253
    communities, 239–246
    of consumption, 22–23
    defined, 17
    of development, 23–24
    of diet, 277–286
    of electricity demand, 98–99
    of energy, 21, 95–103
    of fisheries, 18–19, 229–230
    of forests and wood products, 19–20
    of health and nutrition, 23
    of higher education, 287–290
    of justice and equity, 22
    Kuznets curves and, 24–28
    of manufacturing and industry, 22
    measuring, 18–22
    of population growth, 23
    principles of, 17–28
    of water supplies, 20, 105–117
Sustainable Forestry Initiative (SFI), 207
Sustainable materials, characteristics, 259
Swedish Society for Nature Conservation,
    237
Synergism, 172

Taiga, 203
Tailings pond, 160
Take-make-waste, 256
Talloires Declaration, 288–289, 290, 296
*Tapped* (2010), 117
Tar sands, 80
Temperate forests, 200–201
Thermal expansion of seawater, 66–67
Thermohaline, 64
Total fertility rate (TFR), 49
Toxic air contaminants, 139
Toxins, in soils, 187–188
Trade
    environmental impact of, 27
    global, 26
    U.S.-China, deficit, 81
    global, 26
Transportation, 29
    petroleum, 94
    waste, 148–150
Trash fish, 230, 231
Trash trees, 231

Tree plantations, 20, 201
Triple bottom line, 17, 277, 284
Tropical forests, 197–200
Tropical storms, hurricanes and, 242–243
Trucking, highway safety and, 151
Turf
    adverse effects of lawns and, 136
    allergies and, 136, 138
    defined, 134–137
    land-use changes, 133–134
    lawn care industry, 137
    proliferation, 132–143
Turning Point, 26
Turtle-excluder devices (TEDs), 27

UN Environment Programme (UNEP), 28,
    68, 161
United Egg Producers, 286
United Nations, 31
    Convention on Biodiversity, 298
    Decade of Education for Sustainable
        Development, 289
United States
    electricity demand in, 98–99
    energy development in, 95–103
    gasoline prices, 120, 128
    grain consumption in, 176
    grainland area in, 179–180
    immigration, impact on, 51
    standard of living in, 90–94
    water use in, 110–113
    wind energy development in, 99–101
Urbanization, 23, 185
U.S. Census Bureau, 31, 33
U.S.-China trade deficit, 81
U.S. Commission on Immigration Reform,
    51
U.S. Department of Agriculture, 277
U.S. Department of Energy, 68, 88
U.S. Environmental Protection Agency, 68
U.S. EPA Climate Change, 62
U.S. Forest Service, 20, 303
U.S. Geological Survey (USGS), 157, 211,
    213

Vapor pressure, 163
Veined rapa whelk (*Rapana venosa*),
    221–224
Virginia Institute of Marine Sciences
    (VIMS), 221
Volatile organic compounds (VOCs), 126,
    290

Waste, 109
    categories of, 145
    closing of Fresh Kills landfill, 149–150
    construction and demolition waste
        (C&D), 144–148
    disposal costs, 152
    electronic waste, 152–154
    e-recycling, 154

highway safety and hauling trash, 151
population, 29–31
Supreme Court rulings, 150–151
transporting, 148–150
Waste Management (WM), 155, 266
Waste reduction, 152
Water cycle, 106, 107
Water pollution
agriculture, 114
oxidation of pyrite and, 84
Water supplies
availability, 108
in China, 113–114
climate change, 63
collection of, 108–109
distribution of, 108–109
in Gaza, 115
global use, 105–106

impacts of contaminated, 109–110
Israeli/Palestinian conflict, 114–116
pricing, 117
sustainable, 20, 105–117
in United States, 110–113
value of, 107–108
Weather patterns, 55
West Bank groundwater, 115–116
Wind energy
development in U.S., 99–101
economics, 101
Wind farms, 100
Wind Production Tax Credit (PTC), 101
Wind turbines, 95
Wood products. *See* Forests
World Bank, 28, 82, 110, 129, 167, 191
World Commission on Environment and
Development, 17

World Gold Council, 158
World Health Organization (WHO), 136,
160
World POPClock, 6
World population growth, 31
World proven oil reserves, 70–72
World Resources Institute, 42, 113, 185,
192, 193, 195
World Trade Organization (WTO), 26,
27–28
World Water Council, 105
World Wildlife Fund, 237

Xeriscaping, 138

Zebra mussel, 224–226
Zero power, 4
Zero Waste America, 155